GW01398500

SUN TZU
THE ART OF WAR

BOOK COVER SYMBOLISM
Yellow • Bright Sun, Happiness, Optimism, Hopefulness
Brown • Solid Earth, Stability, Dependability, Reliability

By The Author

LAO TZU'S *TAO TE CHING*
Psychotherapeutic Commentaries
A Wayfaring Counselor's Rendering
of The Tao Virtuosity Experience
[Regent Press, 2016]

CHUANG TZU'S *NEI P'IEN*
Psychotherapeutic Commentaries
A Wayfaring Counselor's Rendering
of The Seven Interior Records
[Regent Press, 2017]

LIEH TZU'S *HSING SHIH SHENG*
Psychotherapeutic Commentaries
A Wayfaring Counselor's Rendering
of The Nature of Real Living
[Regent Press, 2017]

LAO TZU'S *TAO TE CHING*
Sour Journeying Commentaries
A Sojourning Pilgrim's Rendering
of 81 Spirit Soul Passages
[Regent Press, 2018]

TEN ZEN OXEN
Psychospiritual Commentaries
A Way Showing Path of Enlightenment
[Regent Press, 2024]

SUN TZU
THE ART OF WAR

Psychosocial
Commentaries

孫子兵法

A WAYFARING SOJOURNER'S RENDERING
OF
THE WAY OF WARFARING

Raymond Bart Vespe

REGENT PRESS
Berkeley, California

Copyright © 2024 by Ray Vespe

Paperback
ISBN 13: 978-1-58790-692-3
ISBN 10: 1-58790-692-9

E-book:
ISBN 13: 978-1-58790-693-0
ISBN 10: 1-58790-693-7

Library of Congress Cataloging-in-Publication Data

Names: Vespe, Raymond Bart, author.
Title: 880-01 Sun Tzu, the art of war : psychosocial commentaries = Sunzi
  bing fa / Raymond Bart Vespe.
Other titles: 880-02 Sunzi bing fa
Description: Berkeley, California : Regent Press, [2024] | Includes
  bibliographical references. | Summary: "Sun Tzu's "The Art of War" is a
  Chinese classic in military strategy and warfare that has been consulted
  and applied by international leaders for over two-thousand five hundred
  years. It is founded on the fundamental principles, wisdom and practices
  of ancient Chinese Taoism and shares the ethical value of resolving
  conflicts, winning battles and defeating enemies through knowledgeable,
  enlightened and compassionate ways of not fighting a war. "The Art of
  War" is relevant for managing both interpersonal conflicts between
  ourselves and fellow human beings and intrapersonal conflicts between
  our primary executive ego-self and our various subordinate ego-selves.
  Its practical value, relevance, applicability and usefulness are its
  being a context and springboard for encouraging, inspiring and
  stimulating the creating, devising and implementing of your own ways of
  navigating intrapersonal and interpersonal conflicts based upon the
  nature of your unique self, the nature of specific conflicts and the
  nature of particular opponents and adversaries"-- Provided by publisher.

Identifiers: LCCN 2024044657 | ISBN 9781587906923 (trade paperback)
  | ISBN 9781587906930 (ebook)
Subjects: LCSH: 880-03 Sunzi, active 6th century B.C. Sunzi bing fa |
  880-03 Sunzi, active 6th century B.C.--Criticism and interpretation. |
  Psychology, Military. | War--Psychological aspects.
Classification: LCC U101.S96 V47 2025 | DDC 355.02--dc23/
eng/20241017
LC record available at https://lccn.loc.gov/2024044657

Manufactured in the United States of America

REGENT PRESS
www.regentpress.net
regentpress@mindspring.com

CONTENTS

THE WYVERN SYMBOL

The Wyvern is a black military dragon; with wings, two legs and a pointed tail; is associated with combat, war, disease and is strong, powerful and protective; persevering, enduring and benevolent; has immunity to weapons and it fights alongside human beings to combat evil. Wyverns can shape shift into humans.

Caveat

The focus of this book is primarily upon the conflicts and wars of our socially conditioned and determined ego-self rather than upon peaceful resolutions resulting from our being identified *as* enlightened True/Tao-Nature. It is our dualistic and oppositional ego-self that experiences conflicts and engages in wars and not our enlightened non-dual True/Tao-Nature which is an object-free, conflict-free and war-free reality of pure, empty, clear and open consciousness. Therefore, our limited ego-self can be the only 'subject' of the focus of this book and unlimited enlightened True/Tao-Nature cannot be an 'object' of the focus of this book.

Also, this book is not a how-to-do or self-help one that provides numerous actual examples of specific strategic tactics to use for successfully managing, dealing with and resolving both intrapersonal and interpersonal conflicts. Its practical value, relevance, applicability and usefulness are its being a context and springboard for encouraging, inspiring and stimulating the creating, devising and implementing of your own ways of navigating intrapersonal and interpersonal conflicts based upon the nature of your unique self, the nature of specific conflicts and the nature of particular opponents and adversaries.

> The ultimate solution
> to conflict resolution
> is avoiding escalation
> through cooperation,
> through collaboration,
> through mediation,
> through negotiation.

PREAMBLE

There is nothing so detrimental to the positive, optimal and elevated evolution of humanity, human nature, human beings and our universal collective True/Tao-Nature and Higher/Tao-Self than having the energy, force, power and strength of Spiritual knowledge, wisdom and truths either being obscured, obstructed and displaced or being derealized, politicized and weaponized and wielded against human beings by the separate and opposing, judgmental and antagonistic, hateful and hostile, greedy and acquisitive and conflicting and warring false ego-selves of unawakened, unenlightened, unrealized, unactualized, unliberated, undeveloped and unevolved inhumane tyrants and despots; fascist and imperialistic leaders; patriarchs, oligarchs and plutocrats and undemocratic autocrats and dictators who are in positions of power.

❖ ❖ ❖ ❖

Real conflicts and real fights between ego-selves identified as True/Tao-Nature for real reasons of transpersonally defending, safeguarding, preserving, stabilizing, conserving, maintaining, sustaining, regenerating, fulfilling and evolving the viability, dignity, unity, harmony, integrity, energy, vitality, beauty and creativity of universal True/Tao-Nature and complementary Higher/Tao-Selves are more true than unreal resolutions and unreal peace between contradictory and opposed false ego-selves for false reasons of impersonally construing, believing in, satisfyi ng, gratifyi ng, enhancing, maximizing, justifying, aggrandizing and inflating the insubstantiality, artificiality, myth, fiction, fantasy, illusion, delusion, dream and hallucination of socially conditioned, fabricated, contrived and devised false ego-selves.

❖ ❖ ❖ ❖

Currently, there are pervasive conflicts and wars occurring throughout our world. Numerous nations are defensively budgeting, preparing, mobilizing and training for possible or actual armed conflicts and wars at their borders or upon their capitals, sensitive targets and military installations. Some more imperialistic nations are actively readying to take aggressive military actions against and to initiate offensive wars with those other nations. Since we may be on the cusp of WWIII; it is timely for leaders of the global military, governmental and political communities to understand and implement the military strategic tactics of Sun Tzu's *The Art of War* that involve resolving conflicts, defeating enemies and winning wars without fighting them at their great cost of economic resources and their great loss of precious human lives.

All countries love peace.
Most countries hate war.
Many nations fight for peace.
Some nations seem to love war.

All human beings love harmony.
Most human beings hate conflict.
Many people fight for harmony.
Some people seem to love conflict.

MANIFESTO

As sacred human Spirit, blessed human souls and precious human beings; each and every one of us rightfully deserves living and sharing the inborn validity, dignity, integrity, freedom, peacefulness and happiness of real and true human lives that ultimately and intimately are so much more essential, worthwhile, meaningful, significant, important and fulfilling than;

1) murdering, exterminating, eradicating, obliterating, annihilating and 'neutralizing' each other and undergoing suffering, lamenting and grieving over the mass slaughter, carnage, 'collateral damage', destruction, demolition and devastation brought about by and in wars and

2) experiencing, undergoing and enduring the intense feelings of anxiety, anger, depression, fear, panic, hopelessness, helplessness, pain, anguish, despair, hatred, horror and terror over the threat, prospect, presence and conduct of ongoing wars.

There is no greater or more utterly horrible misfortune, tragedy, disaster, calamity, catastrophe and atrocity of and for humanity than to have fellow human beings and kindred human souls conflicting and contending with, attacking, fighting, battling, killing, slaughtering and massacring one another in wars that now appear to be approaching cataclysmic and apocalyptic proportion.

There is no greater or more utterly terrible failure, decay, devolution, destruction, desecration and loss of and for humanity than to have fellow human beings and kindred human souls injuring, wounding, disabling, crippling, maiming, defeating, conquering, vanquishing and destroying one another in wars that now appear to be approaching epic and pandemic proportion.

There is no greater or more utterly unviable limitation, deficiency, deviation, obstruction, hindrance, disruption, error and futility of and for humanity than to have fellow human beings and kindred human souls promoting, resorting to, initiating, conducting, engaging in, perpetuating and winning offensive (in both senses of the word) wars that are absolutely, ultimately, essentially, fundamentally, completely and humanely needless, senseless, aimless, pointless, purposeless, fruitless and meaningless.

It is sad and discouraging that such a large part of the planetary and global evolution and the governance and economy of our human culture, countries, nations and society involves and focuses upon better learning and knowing how to defend against possible and real enemies; to conduct, manage and win wars and to develop, perfect, manufacture, stockpile and employ the best and most powerful state-of-the-art destructive weaponry.

What is currently going on in our world with regard to conflicts and wars is an unfortunate, tragic, deplorable and lamentable reality and is not simply an objectified and informative 'narrative' reported on by 'talking-head' newscasters, analyzed by social media commentators and received by a general public too accustomed to spectatorially viewing and vicariously identifying with real people and actual events that are present and occurring in real time.

As unsettling and unnerving, upsetting and gut-wrenching, head-spinning and heart-breaking as they are; my hope is that the increasing frequency of live real-time media presentations and news pictures:

1) of the military operations, injuries, deaths, destruction and losses involved in current wars, e.g., air strikes, explosions, missile interceptions and fighting soldiers; screaming women, crying men and bleeding children; human beings carried on stretchers; orderly arranged body bags on the ground and mass graves; destroyed homes, buildings and incinerated vehicles; alive and deceased human beings being lifted out of rubble, etc.;

2) of displaced parents, children and infants huddled together on the ground in refugee camps alongside the few belongings

that they could take with them; of starving children scraping bits of food from empty pots and bowls and of mass crowds of hungry and sick human beings fighting and crawling over each other to grab food and medications from humanitarian aid airdrops and distributions;

3) of the live moving testimonial interviews with human beings who are personally, sadly and deeply emotionally experiencing the disastrous effects, destructive and devastating results and extreme deprivations and losses involved in military war operations, genocides, war crimes, armed conflicts, gun violence, homicides, suicides, crimes, etc.;

4) of police activity coverage, body camera footage and closed-circuit surveillance videos showing crimes being committed; pursuits and crashes occurring; violence, robberies, shootings, woundings and killings happening; police apprehendings, subduings, handcuffing and arresting being done often with abuse, excessive force and brutality, etc.;

5) will promote and facilitate accuracy, comprehensiveness, representativeness, integrity and honesty in journalistic coverage and media reporting that will not involve traumatizing, overwhelming and excessively stimulating viewers; editorial censorship, selectivity and bias; misinformation, disinformation and propagandizing, etc. and will also include the many balancing positive events occurring daily in our nation and the world;

6) will positively impact and serve to poignantly awaken us to, and raise our consciousness about, the inhumane, unconscionable and absurd tragedy, atrocity, horror and terror of wars and the increasing frequency of abuses of power, racially and ethnically-motivated hate crimes, domestic terrorism, mass shootings, gang violence, homicides, suicides, systemic racism, dysfunctionalities and injustices, etc, in our society and

7) will personally prompt and compassionately compel some concrete, direct, relevant and potentially significant and meaningful 'off-couch' action by us; even if it is lighting a candle

and saying a prayer for, or sending positive and healing energy to, those unfortunate human beings; civilian and military and victims, survivors and aggressors on both 'sides' of wars, armed conflicts and violent crimes; who regrettably have experienced or are currently experiencing unforgivable inhuman atrocities and unforgettable inhumane realities and who are equally deserving of our absolute, ultimate, universal and essential human birthright to deeply experience and to fully share a real and true human life of light, peace, freedom, intimacy and happiness.

Wars . . .
are not to be willed.
Human beings . . .
are not to be killed.

Blood . . .
is not to be spilled.
Human lives . . .
are to be fulfilled.

DEDICATION

This book is dedicated; with grief, compassion and gratitude; to the living memory and surviving kin of those human beings who have perished fighting in wars and to those human beings who are currently risking their lives protecting and defending the lives, safety and security of fellow human beings by fighting in wars; who are casualties of and who have been wounded and injured in wars and who are suffering from post-traumatic stress disorders as a result of being involved in wars.

This book is dedicated; with care, concern and hope;

1) to the children who are frightened, traumatized and displaced by living in, or forced to emigrate from, war-torn countries and who are suffering their resultant injury and the injury, death and loss of family members and loved ones and the disruption of their families;

2) to the children who are confused, upset, frightened and traumatized by witnessing the armed conflict, injury, death and destruction broadcast in TV news and social media coverage and graphic real-time portrayal of wars;

3) to their mothers, fathers, siblings, grandparents, relatives, peers and friends who have perished because of wars and those who are still alive and able to offer them and to share some modicum of safety, care, support, empathy and understanding.

This book is dedicated; with reassurance, happiness and gratitude; to the leaders, governments, organizations, personnel and citizens of democratic countries and nations who/that are providing humanitarian support, aid and assistance to displaced refugees.

This book is dedicated; with regard, respect and gratitude;

1) to past and present military medics, physicians without

borders, paramedic, emergency medical, hospital and clinic personnel; police, fire and first responders; aid workers, volunteers, selfless unsung heroines and heroes and good samaritans and the many American Red Cross workers (since 1881) who have courageously used, risked, injured and may have unfortunately lost their lives coming to the rescue and emergency aid of, and providing disaster relief for, fellow human beings;

2) to past and present emergency (911) and suicide/crisis/ health issue (988) hot-line staff members who daily have used or are now using their skills to assist fellow human beings in dealing with themselves and the issues, crises and emergencies of their lives;

3) to past and present dedicated surgeons, physicians, nurses, physician assistants, aides and medical staff who daily are devoted to repairing bodily injuries; healing diseases, illnesses and medical conditions; keeping fellow human beings whole and healthy and preserving and extending human lives;

4) to past and present pysychoanalysts, psychiatrists, clinical psychologists and social workers, psychotherapists, counselors, various therapists and clergy who are daily involved in assisting fellow human beings in dealing with and resolving psychological, interpersonal, social, economic, vocational and spiritual, etc. issues and conflicts;

5) to the commitment of all of the above to protect the lives, safety, security, peace and physical, mental and emotional health and well-being of human beings and

6) to their particularly dealing with emergencies and crises that constitute threats and dangers to themselves and to fellow human beings and that involve the causative and resultant ways in which we are in conflict and at war with ourselves and each other.

This book is dedicated; with inspiration, comradery and encouragement to each and every awake and consciously aware, law-abiding, peace-loving and courageous human being;

1) who goes about daily living, hallowing, treasuring,

cherishing, experiencing and sharing our sacred, blessed and precious human life in ways that are dignified, respectful, humane, civil, circumspect, inclusive, empathic, compassionate, intimate, meaningful and fulfilling and

2) who is presently dedicated, devoted and committed to; and actively engaged in and deeply involved in; some significant ways of non-aggressively confronting, addressing, negotiating and resolving conflicts and protesting, preventing, stopping and ending wars.

A grateful dedication
to all human beings
supporting creation,
fostering transformation,
protecting our people,
defending our nation,
sustaining its station
with avid determination,
insuring its continuation,
for a lengthy duration.

A grateful appreciation
of all human beings
making the decision,
to serve the mission
of uniting our country,
healing our division,
crafting our revision,
with utmost precision
avoiding any collision,
or international derision.

ACKNOWLEDGEMENT

A Grateful and Appreciative Acknowledging:

Of ancient Chinese Tao-Masters, Lao Tzu, Chuang Tzu and Lieh Tzu who; during the civil strife, hegemonic conflicts and internecine wars between feudal states and the great human carnage closely preceding and during the Warring States period of Chinese history; strongly deplored tyrannical autocratic rulership, the rampant waging of bloody war and the resultant mass slaughtering of human beings.

Of ancient Master Sun Wu Tzu who authored *The Art of War* and provided national and international leaders with knowledge, strategies, tactics and maneuvers enabling the winning of wars quickly, defeating enemies without fighting, conserving economic resources and saving the precious lives and blessed souls of human beings.

Of my wise, compassionate and skillful Hindu, Taoist and Buddhist teachers and mentors and our focus more upon reality, truth, peace, freedom and happiness rather than upon unreality, untruth, war, bondage and despair.

Of my awesome, amazing and astonishing bi-generational daughters Arianna and Cheryl and their families for their long-standing and ongoing acceptance, love, caring, kindness, understanding, support and assistance and their disinterest in, and the virtual absence of, serious disagreements, conflicts and fights between them and with me.

Of my wonderful and delightful eight adult, young adult and young grandchildren who have shown and taught me that differing, disagreeing, disputing, conflicting and fighting can be part of close relationships and can result in greater growth, personal power, individuation, freedom, happiness, peacefulness, connection, empathy and intimacy.

Of my nine former graduate school psychology students, wonderful human beings, beautiful souls and now close friends;

Roger M. and Susan W., Bob H, Ted C., Art R., Barry S., Richard N. and Richard and Helen D., whose continuing heart-felt love and friendship I profoundly value and respectfully appreciate and whose continuing and deepening personal sharing, like that of my close family members, gives me and my life extraordinary, incredible and inexplicable enjoyment, comfort, meaning, happiness and fulfillment.

Of Mark Weiman of Regent Press and his usually unusual and ordinarily extraordinary formatting of this book and his prioritizing and expediting its crafting so that it did not have to be published posthumously without my seeing it and being able to give copies to family members, friends, interested folks, bookstores and libraries.

Each and every one of you is with me each and every day.

A grateful acknowledgement:
Of teachers and masters
who shunned all wars
and human disasters,
united contrasters,
became long-lasters.
Of every human being
committed to peace,
caring and freeing,
happiness and seeing,
not I-ing but we-ing.
Of family and friends
resolving any conflicts,
achieving harmonious ends,
knowing softness lives and bends,
loving kindness heals and mends.

Recognition

A Thoughtful and Reflective Recognizing:

Of the continuing prevalence of recorded territorial, national and international disagreements, disputes, hostilities, conflicts and wars that have historically occurred in our world from 3100 BCE until the present time; in spite of over twelve-thousand three-hundred various peace edicts, contracts, covenants, agreements, accords, pacts, treaties, resolutions and declarations having to do with boundaries, rights and claims; cedings, accessions, annexations and grants; sovereignty and independence; neutrality, alliances and unions; reciprocal assistance, trade relationships and economic cooperation; renouncing war as an instrument of national policy, establishing rules and crimes in the conduct of war and limiting the proliferation and stockpiling of nuclear armaments and arsenals.

Of what appear to be ongoing pervasive world-wide hostilities, conflicts and wars of epidemic and pandemic proportion that are initiated, perpetrated, conducted and perpetuated by:

1) power-hungry and war-mongering tyrannical, despotic, fascistic, monarchistic, militaristic, autocratic, dictatorial and patriarchal 'leaders' and their governments that hold and implement nationalistic and isolationist ideologies and foreign policies, agendas and objectives that result in conflicts and wars that destabilize and imbalance regional geopolitics, disrupt national and international harmony, ruin the lives of human beings at home and end peaceful co-existing and cooperative relationships with neighboring countries, nations and peoples;

2) the existence, propagandizing, disinformation, strategies, tactics and operations of various terrorist and extremist organizations, e.g., using human beings as suicide bombers and

shields; conducting raids, rampages and massacres; slaughtering women, children, men and elders; torturing hostages; raping women, burning children and beheading human beings and

3) that result in disastrous national, international and global community ramifications, consequences and repercussions throughout our sacred world and precious planet.

Of increasing firearms purchasing, owning and carrying; gun violence, suicides, homicides and accidents; police-involved shootings and misuse of deadly force and citizen-used stand your ground law; domestic terrorism and the use of assault weapons in racially and ethnically-motivated hate crimes, targeted and random shootings and mass murders in our communities that also are of epidemic proportion and occurring nearly every day in our country.

Of the alarming and sad reality that far too many (and one is too many!) objections, differences, disagreements, disputes, arguments, conflicts, bar-room brawls, road rages, auto accidents and physical fights are now being 'settled' and ended by one party physically injuring, shooting, wounding and/or killing the other.

Of the senseless killing of human beings who were shot to death for making too much noise in a neighborhood, infringing on the property boundary of a neighbor, objecting to a neighbor's dog barking or loud music, accidentally ringing the doorbell of a wrong house, being mistaken for a 'porch-pirate', inadvertently turning around in someone's driveway, etc..

Of all human beings who are sincerely committed to and deeply and fully engaged in, dealing with, settling, resolving, reconciling and ending differences, oppositions, disagreements, arguments, disputes, conflicts, contentions, fights, battles and wars through non-hostile, non-aggressive, non-violent, equitable, reciprocal and peaceful means of statespersonship, mediation, diplomacy, negotiation, compromise and mutual agreement.

Of all human beings as they honor, protect, regard, respect, treasure and cherish fellow human beings as sacred and blessed beings, beautiful sisters and brothers, precious and kindred souls and embodied Spirit and who are not to be endangered, exploited, trafficked, hated, terrorized, traumatized, bullied, intimidated, molested, abused, violated, harmed, injured, maimed, crippled, disabled, raped or killed and their bodies dismembered, incinerated, buried in forests, discarded by roadsides and in rivers or stuffed in suitcases and trash bags and disposed of in dumpsters and city garbage facilities.

Of human beings who are victims and survivors of incest, rape and pedophilia; sexual violence, abuse, molestation, exploitation, prostitution, 'grooming', voyeurism, stalking, groping and harassment who have come forward with allegations of sexual misconduct by identified perpetrators and who have too often been ignored, discounted and blamed for their experiences; ignoring the reality that the defending claim made by aggressive and forceful perpetrators of 'consensual agreement' and 'consent' on the part of victims masks the latters' underlying trauma and fear-based silence, paralysis, passivity and compliance rather than evidences a willing agreement and participation in the activities.

Of innocent bystanders randomly caught, injured or killed in the cross-fire of gun violence, gang wars and shoot-outs between police and suspected or identified criminals.

Of children who have accidentally died through having had access to, finding, carrying, playing with and using firearms that were not emptied of ammunition, safely locked up or properly stored out of their sight and reach.

Of children, adolescents, adults and elders who are afraid to be in parks and playgrounds, schools and colleges, houses of worship, medical and care facilities, shopping malls and stores, restaurants and bars; to be at public community events, concerts, large gatherings and outdoor family celebrations and to traverse, stroll their infant children, walk their dogs and jog or cycle on city and neighborhood streets and paths.

Of the weaponizing of governmental operations and politics,

partisan infighting and the apparent legislative deadlock regarding necessary actions to be taken to ban the manufacture of assault weapons, limit the production of firearms in general, restrict their sales, control their ownership and carrying and to deal with the underlying mental health issues often involved in their accessibility, acquisition, accumulation and use in violent crimes, mass shootings, etc..

Of human beings who have been or are being unfortunate victims and courageous survivors of physical, sexual, mental, emotional, psychological and/or verbal abuse; and domestic violence or who have been or are being in any way and form or to any degree and extent; traumatized, abandoned, invalidated, neglected, targeted, oppressed, preyed upon, exploited, victimized, bullied and terrorized.

Of human beings who have been deceived, duped, cheated, betrayed, robbed, kidnapped, abducted, ransomed and exploited in predatory and parasitic relationships by desire-, status-, power-, profit and wealth-seeking quasi-human beings who capitalize upon the needs of human beings for food, water, shelter, safety, health care, money, security, stress-reduction, pain-relief, entertainment, peace, freedom, happiness, something or someone to trust, to believe in, to be loved by, etc. and their anxieties and depression over and fears of the lack thereof.

Of human beings who have been or are being objects of prejudice, chauvinism, misogynism, racism, sexism, ageism, xenophobia and hate crimes; systemically discriminated against; persecuted, socially and economically segregated; racially profiled, oppressed, dominated, subjugated, dehumanized, depersonalized, negated, disempowered, disenfranchised, disadvantaged, underrepresented, stigmatized, marginalized, pathologized, criminalized, patronized, infantilized, victimized and who are needy, suffering, struggling, toiling, destitute, poor and homeless.

Of human beings who are suffering malnutrition, starvation, disease, illness, deprivation, impoverishment and humanitarian crises due to civil wars, extremist groups and terrorist activities in their countries and of those who have been forced to evacuate their homes and to migrate from their native homelands and to seek safety, refuge and asylum in foreign countries due to crimes, dangers, conflicts, violence, atrocities and wars in their own countries.

Of human beings who daily exist censored, falsely accused, unfairly tried, unduly punished and unjustly imprisoned; in bondage, hopeless and helpless despair and fear for their safety and lives and who are unable to experience and to enjoy any degree of well-being, peace, freedom and happiness due to 'living' in totalitarian countries corruptly, tyrannically, oppressively, abusively and harshly ruled by despotic, monarchistic, autocratic, militaristic, dictatorial, fascistic, controlling, dominating, subjugating and enslaving patriarchal 'leaders'.

Of the normal responses and reactions of those human beings whose freedom, independence and autonomy are being restricted, constricted, restrained and constrained that, from the standpoint of the autocracy, are regarded as resistance, disobedience, defiance, rebellion, revolution, insurgence, insurrection, betrayal and treason and which are only further and more harshly controlled, disciplined, penalized, punished, thwarted and ultimately quashed.

Of the reality that national and international wars and conflicts are primarily due to the nationalistic, isolated, untransparent, irresponsible, unaccountably-held and corrupt autocratic, patriarchal and dictatorial leadership and governments of countries and nations and not due to the will, choice and preferences of their people; e.g., as evidenced by prevalent non-discriminatory, cooperative and collaborative relationships between individual or groups and teams of ambassadors, envoys, diplomats and professional artists, scholars, scientists, researchers, educators, journalists, athletes, astronauts, students, tourists, children, etc. who do not have animosity toward each other.

Of the principal source of war being sovereign and autocratic presidential leadership by the one patriarchal 'man at the top'; and his tightly-knit regime of officials, military advisors, lieutenants, generals, cronies, henchmen and puppets; rather than by the democratic leadership of governing bodies that are more reflective of the realities, needs, benefit and will of 'the people' through administrative, legislative and judicial councils, cabinets, parliaments, assemblies and effectively working bi-partisan (ultimately non-partisan) congresses, senates, houses of representatives, etc..

Of even the governmental leadership of some democratic nations that can, may, and do support, declare and/or engage in cold, proxy and hot wars with other nations; ignoring, disregarding, overriding and acting contrary to the essentially, fundamentally and humanely peaceful collective will of the majority of its people.

Of 1) the judicial branches of the governments of countries and nations where judges are nominated and appointed by partisan leaders and who also are partisan in their political identification, opinions and decisions rather than being truly, objectively and fairly non-'judging' and able to render non-partisan and unbiased rulings and

2) the boards of directors of conglomerates and corporations; originally charged with the responsibility of overseeing ethical, responsible and accountable business conduct and equal, fair and humane employee rights; that are often overly aligned with CEOs, COOs and CFOs and their coteries, military-like policies and practices and army-like hierarchical organizational ranking and regimentation and are, therefore, disempowered and unable to fulfill their original purpose of benefitting loyal, dedicated, hard-working and productive employees rather than satisfying financial investors and shareholders.

Of the unfortunate side-effects of a capitalistic, collectivist, corporatist, commercialist, consumerist and competition-based society, profit-over-people-based economy and militaristic war-like mentality; rather than a community (not communist!), partnership and cooperation-based society, nonprofit and people-over-profit based economy and peace-like mentality that does not involve the often ruthless, cut-throat, deceptive, unethical and corrupt objectives, policies and practices of the former.

Of the sad reality that some of the positive discoveries, creative inventions and constructive technological advances that are made also become co-opted for negative purposes and uses by 'negative' human beings, groups, institutions, nations and governments who/that are driven by ego-centric needs for power, money, control and dominance and motives of security, greed, hatred and fear, e.g., atomic energy, lasers, drones, computers, social media, robots and artificial intelligence.

Of the hundreds of billions of dollars of 'defense' contracts, budgetary allocations and expenditures of the 'military-industrial-complex'; a nexus of government, politicians, armed forces and the defense industry and its dangerous potential for allowing mutually vested interests, conflicts of interest and increased business opportunities for the private for-profit overproduction of military weaponry and technology that can be a misplaced national focus and power, a strong determinant of public and foreign policy and a seriously posed threat to democratic society.

Of various other similar mega profit-driven 'industrial' complexes of the government, lobbying politicians and private sector companies that relate to human being, health, rights and life; that involve creating, determining, inflating, over commoditizing, commercializing and capitalizing on human needs and social values and that are socially detrimental and negatively impact the quality of human life.

Of the organic and food, medical and pharmaceutical, evangelical and educational, banking and finance, insurance and real estate, legal and prison, data and technology, 'green' and solar, automotive and transportation, gas and oil, vacation and tourist, tobacco and alcohol, sports and entertainment, wedding and funeral industrial complexes as examples of the above.

Of the unfortunate reality that the original intent and purpose of a bi-partisan political system that was one of inclusion, cooperation, checks-and-balances, agreement and ultimate unity; like most all duality-based things; seldom reaches the unanimity of a trans-partisan unity and becomes bogged down in, and never moves through and beyond, chaotic dualistically-divided, unilateral and mutually exclusive positions; one-sided disagreements; war-like power struggles and conflicts, dysfunctional deadlocks and impasses, narrow majority rule and minority neglect, vetoes and filibusters, etc..

Of the seeming reality that there is little room left 'under the bus' after so many 'others' have been 'thrown' there by denying and unaccountable political leaders blaming others and that there is little space left 'down the road' after so many 'cans' have been 'kicked' there by defying and irresponsible political leaders avoiding issues.

Of the only relative effectiveness and partial successfulness; but the apparent ultimate powerlessness and limited direct and conclusive intervening; of national and international organizations such as The United Nations (UN) and its Security, Human, Women's and Child Rights Councils; The World Health Organization (WHO), The Center for Disease Control and Prevention (CDC), The United States Department of Health and Human Services (HHS); The National Institute of Health (NIH);The International Court of Justice (ICJ) and Criminal Court (ICC); Amnesty International (AI) and Human Rights Watch (HRW); the United Nations High Commissioner for Refugees (UNHCR), International Children's Emergency Fund (UNICEF), Educational, Scientific and Cultural Organization (UNESCO) and, to a lesser degree, the North Atlantic Treaty Organization (NATO) and UN sponsored peace-keeping and refugee-assisting forces from various nations.

Of the existence of charters, by-laws, purposes and visions

of these organizations purported to provide various services that insure, improve and enhance the safety, health, well-being, welfare, security, culture, dignity, respect, rights, justice, equality, cooperation, solidarity and peace of human beings and nations and to protect them from threats, violations, abuses and the effects of disease, illness, disasters and wars, etc..

Of the reality that violations of the objectives of the above organizations do not appear to be expeditiously acted upon in crucial, substantial, meaningful, significant, important and conclusive societal, national and world-changing ways that vitally and deeply matter to human beings, e.g., doing little or nothing beyond identifying; but not quickly charging, prosecuting and trying; the principals of the continuing inhumane operations of well known and closely monitored fascistic governments, terrorist and extremist organizations, drug cartels, groups and gangs who/that are actively committing war crimes, engaged in unlawful, illegal and fraudulent criminal activities.

Of some of the above organizations' questionable and problematic domestic and foreign policies, international diplomatic relationships, government interests and conflicts of interest, proxy and shadow funding, dark money, failure to act against corruption and fraud, systemic investigatory biases in research and drug trials, harmful and fatal medication and vaccination side-effects, etc..

Of the charters and by-laws of the above organizations apparently not including immediately, absolutely, directly, concretely and substantially intervening in any and every essential, necessary, possible, practical and conclusive way to deal with the deeper root causes of war and disease inherent in outdated, outmoded, unviable and problematic models of leadership, government and health care and to prohibit and stop the wars, genocide and crimes against humanity and to prevent and cure the diseases, illnesses and sicknesses of human beings.

Of the over some seventy years that a number of the above

organizations, e.g., the United Nations; while informative and to some extent helpful in dealing with the effects of war in certain necessary, significant, important and consciousness-raising ways; seem to have primarily:

1) gathered, processed, disseminated and announced relevant, useful and informative statistical facts and data regarding the reality, course and spread of wars and diseases;

2) researched, observed, monitored and reported risk lists of the kinds of events and incidences and their locations, frequencies, degrees and extents; etiologies, demographics and epidemiologies and morbidity and mortality rates, deceased and injured body counts;

3) periodically convened to reflect upon and debate issues; make, vote on, pass and veto resolutions; make acknowledgements; provide rhetorical advisory opinions and recommendations; issue periodic alerts and warnings; consider preventative measures and make calls for humanitarian pauses during wars and opening humanitarian borders and corridors for necessary provisions, disaster relief and safe evacuations of civilians;

4) investigated, observed, monitored and, at times, issued orders and petitions for warring factions to cease or pause fire, reduce civilian casualties and not disrupt regional stability and balance by changing the conduct of wars and/or disarming and ending wars involving crimes against humanity and violations of international humanitarian law and the rules of war which, however, are largely ignored by the warring parties and rarely enforced and followed through with;

5) targeted and exposed national and international human rights violations and non-compliance with accords made by offending leaders, governments, organizations, groups and individuals and defended and have secured justice for human beings who are victims, e.g., of discrimination and apartheid; religious and political unfreedom; abduction, sexual exploitation and trafficking; arbitrary arrest and detention; inhumane treatment,

torture and civilian killing in wartime; political executions and assassinations, etc.;

6) appealed to the moral conscience of leaders, countries, nations, organizations and groups committing atrocities of wars and have supported, assisted and contributed to empowering affected human beings in their efforts to deal with and to counteract human rights and civil liberties abuses and violations;

7) occasionally requested funding and programs for countries to provide for the emergency assistance and relief of humanitarian crises and refugees, however, a significant portion of which can often be diverted to administrative overhead 'costs' rather than used for the direct benefit of the intended people affected;

8) been criticized for having an unrepresentative, undemocratic and unequal systemic organizational structure that favors the interests of a few super-power member nations with permanent seats and veto powers, e.g., the UN Security Council, and supercedes the politics, interests and voting power of the majority of less economically developed and neo-colonial member nations, e.g., African nations, in the UN General Assembly.

9) out of a seeming respect for the sovereignty and self-determination of nations (to also control and endanger their people and eliminate and annihilate enemies?) and to demonstrate some kind of non-interfering 'political correctness'; themselves become organizational microcosmic disparate embodiments and reflections of the international leaderships, power-differentials, conflicts and wars that they are attempting to democratize, unite, resolve and/or end.

Of the Institute for the Study of War (ISW) that is concerned with diplomatic, philosophical, psychological, social, political and economic aspects of war. It researches, analyzes and provides assessments and informative reports and updates of current international global issues of defense and foreign affairs and ongoing military operations and insurgent attacks that improve multi-national abilities to respond to emergent threats to national security and to execute necessary military operations to achieve strategic objectives. Critics focus on the organization's generally aggressive and 'hawkish' foreign policies.

Of NATO and its Response Force (NRF) as the one body of North American and European democratic member countries and nations comprising a basically military alliance that:

1) is collectively led, directed and operated by a council, high-level ambassadors and command operations committees and joint readiness and multi-national crisis response and full combat task forces primarily led by international leaders and high-ranking military generals and admirals.

2) is instituted to safeguard the rule of law, rights, freedom, security and peace of human beings and all democratic member countries and nations by political and military means.

3) is designed to collectively and cooperatively protect and defend any member country and nation from, and/or to retaliate, attacks by non-member countries and nations and terrorists and threats of nuclear wars, the use of weapons of mass destruction, cyber-attacks, etc..

4) is a global network that is also available as a collaborative military resource for non-member countries and nations engaged in armed conflicts and wars, e.g., by providing financial, arms, supplies, training and logistical support.

5) currently functions autonomously and independent of its former need for authorization of military operations and strikes to be granted by the UN Security Council.

6) conducts military trainings and exercises except in those member countries and nations electing to prohibit military bases, presence and activities and nuclear warheads.

7) standardizes military equipment, policies and procedures; determines the stationing and deployment of military forces and enforces resolutions such as arms control, nuclear deterrence and no-fly zones.

8) does not address, deal with, intervene, participate and

assist in any civil wars that may be occurring in member countries and nations.

9) conducts aerial surveillance and reconnaissance operations over countries and nations that are involved in conflicts that could potentially escalate into wars.

10) deploys ground, maritime and air forces when military interventions are necessary in member countries and nations that are attacked by and at war with other non-member countries and nations.

11) has assisted evacuating civilians, delivered humanitarian aid and emergency relief supplies and successfully conducted a number of effective and decisive military missions, campaigns and operations in member countries and nations and throughout the world.

12) has been criticized for being primarily funded by the United States and protecting, implementing and defending its interests; for some pre-emptive military interventions and questionable aggressive operations; for expanding and becoming a global police-force and for risking increasing a global militarizing of the planet.

Of all governmental and military organizations, e.g., The Department of Defense (DOD) and 1) its Defense Intelligence Agency (DIA), Defense Counterintelligence Security Agency (DCSA) and Defense Security Cooperation Agency (DSCA); 2) its Branch Departments of the Army, Air Force, Navy, Marine Corps, Space Force, Coast Guard and National Guard; 3) its eleven Unified Combatant Command Branches, e.g., the Special Operations Command (SOCM) and 4) its elite Special Operations Forces (SOF) such as the Army Green Berets, Delta Force and Rangers (to a degree); the Navy Sea, Air, Land Teams (SEALS) aka Frogmen; the Marine Corps Raiders (MARSOC) and the Air Force Combat Controllers (CCT):

1) that comprise the largest organizational numbers of military personnel, facilities, bases, vehicles, equipment and

armaments and the highest budgeted funding and expenditures of the United States government.

2) that provide military forces to deter war; protect and ensure our nation's safety and security; defend our nation from threats and attacks and defeat adversaries, terrorists and enemies;

3) that perform national and international peacekeeping, disaster response, arms control, nuclear deterrence and law enforcement duties and activities;

4) that involve operating in all kinds of terrain, under all types of extreme conditions and in all parts of the world, e.g., to capture and assassinate 'high value targets', enemy personnel and terrorists and conduct covert special reconnaissance, sur- veillance and counter-insurgency missions; raids, infiltrations and extractions, direct strikes, hostage rescues, etc. and

5) that rigorously train highly qualified military personnel to conduct high-risk missions and operations using unconven- tional and guerrilla warfare methods.

Of the failure to confront, engage, arrest, charge, pros- ecute, try, convict and sentence 1) the leaders and heads of those often governmentally well known extremist and terrorist organizations, cartels, gangs and groups who/that are disrupt- ing democratic society by controlling, threatening, terrorizing, traumatizing, exploiting, addicting, trafficking, injuring and murdering human beings and 2) the foreign and domestic own- ers, investors and operators that are involved in commercially growing, manufacturing, transporting, smuggling, distributing, black-marketing and selling illegal, dangerous, harmful and potentially lethal drugs and various counterfeit prescription medications laced with them.

Of those leaders and governments of countries and nations who/that are not being held responsible and accountable for;

1) violating international humanitarian law and the rules of what is and is not allowed in war, e.g., no cluster bombs, exploding bullets, asphyxiating and poisonous gases, chemical

and bio-weapons, blinding lasers, land mines, nuclear weapons and civilian attacks and

2) perpetrating crimes against humanity, e.g., murder, massacres, extermination and atrocities, enslavement and deportation; mass systemic rape and other inhumane acts; political, racial and religious persecution; genocide and 'ethnic cleansing' (reminiscent of racial superiority, master race, white supremacy and eugenics doctrines) all of which are currently going on; unabated, unconfronted, unaddressed, undealt with and undisciplined in our world.

Of the executive, legislative and judicial governmental bodies of various countries and nations that do not appear to be addressing real, compelling, critical and urgent issues with regard to the safety, health, security, dignity, equality and peace of human beings throughout the world; the human rights to which are especially threatened, endangered and violated in nations absolutely ruled by fascistic plutocrats, oligarchs, tyrants, despots, patriarchs, autocrats and dictators.

Of the National Security Agency (NSA), Central Intelligence Agency (CIA), Defense Intelligence Agency (DIA), Secret Service Agency (SSA), the Department of Justice (DOJ), the Federal Bureau of Investigation (FBI), the Department of Homeland Security (DHS), the Drug Enforcement Administration (DEA) and the White House Police Force;

1) as the first and front lines of national and international defense and federal law enforcement to keep the country safe, insure domestic security and freedom, protect civil rights and safeguard United States leaders and the American people from threats, harm and crimes and that strongly influence and determine governmental, domestic and military policy-making and program decisions regarding national safety and security and their violations, requirements, issues and concerns;

2) for providing national disaster response and protection from business fraud and counterfeit goods and national safety

and security threats to the climate, economy and borders and from terrorist and gang activity, child exploitation, illegal immigration, human trafficking, narcotic and illicit drug smuggling, illegal weapon exports, organized crime, etc.;

3) as the SSA, has secret service agents who protect and ensure the safety of political leaders and personnel and their families and visiting international political leaders and heads of state;

4) as the DOD; United States Space Force (USSF), Strategic Command and Command Force; NSA and DIA and the National Aeronautics and Space Administration (NASA); variously explores space; records and investigates sightings, reports and activities of unidentified flying objects (UFOs), unidentified aerial/anomalous phenomena (UAPs) and near-earth asteroids as constituting possible credible threats to the planet and nation and performs global airspace surveillance, warning, command and defense; nuclear deterrence and strike readiness in the event of potential and actual attacks;

5) as the FBI and CIA; investigates and deals with domestic and foreign cyber-hacking and cyber-attacks; domestic crimes involving citizens and foreign nationals, workers and immigrants and racially motivated hatred, violence, unruly demonstrations and protests due to, e.g., anti-Semitic, Islamophobic and other xenophobic ideologies;

6) as the CIA; is a civilian domestic and foreign intelligence service and espionage agency of the federal government that globally monitors, obtains, collects, processes and analyzes objective intelligence, signals intelligence and 'mines' cell-phone data and media and web-site meta-data to discover and pre-empt domestic and international threats and cyber-threats and to further national safety and security needs, interests, concerns and objectives;

7) as the CIA; conducts secret, covert, clandestine, surreptitious, spying and espionage operations reportedly involving hacking, eavesdropping and 'bugging; secret, 'black', shadow and false-flag strategic operations; tactically 'staged' events,

planted evidence and intentional false propaganda, subterfuge, deception, misinformation and disinformation;

8) as the CIA; gathers intelligence information about the locations, intentions, capabilities and actions of foreign adversaries and enemies in order to gain decisive advantage over or defeat them through a global network of intelligence and counter-intelligence operations, using spies and double espionage agents and planning and conducting raids, extractions, captures and assassinations by special operations military personnel;

9) as the CIA; has an independent, extra-governmental and paramilitary Special Activities Division (SAD)/Special Operations Group (SOG) that is one of the most elite, rigorously trained and highly skilled clandestine special forces and strategic tactical teams conducting various covert operations involving land, marine and aerial surveillance and reconnaissance; counter-intelligence, counter-insurgent, anti-terrorist and hostage rescue operations; insertions, infiltrations and extractions behind enemy lines; raids and demolitions and direct combat missions and is primarily comprised of former elite tier one special operations military personnel from the Army Green Berets, Delta Force and Rangers; Navy Seals and Marine Corps Raiders;

10) as the CIA; has historically primarily focused upon the activities of international cold war anti-Democratic, Socialist and Communist nations, reportedly has maintained assassination manuals and targeted 'disposal lists' and has trained personnel in the most efficient and effective assassination methods and ways of removing and 'neutralizing' leaders that are ostensibly deemed to pose threats to our U.S. democracy, national sovereignty and interests, economic stability and prosperity, international leadership, superiority and supremacy, etc.;

11) as the CIA; reportedly has conducted various questionable overt and covert human experiments involving sensory deprivation, mind-control, brainwashing, hypnosis, use of psychoactive substances and hallucinogens, personality

deprogramming and repatterning, moral character obliteration, cancer induction, the use of biological weapons and germ warfare and the contamination of food and water sources with the purpose of discovering the best methods and quickest ways of harming, sickening, poisoning, killing and assassinating individual human beings, groups and whole populations;

12) as the CIA; reportedly has experimented with uninformed and misinformed human research subjects under false and deceptive circumstances whereby many of the results were disastrous and many of the subjects incurred and suffered lifelong trauma, mental illness, psychoses, psychiatric hospitalization and institutionalization and who often went missing, disappeared, died or committed suicide and

13) as the CIA; has generated a host of allegations and various conspiracy theories about, e.g., serving military objectives and making war appear retaliatory and justifiable through deception about unprovoked attacks; torturing detainees with 'enhanced interrogation techniques'; blackmailings and entrapments, subversion, sabotage and assassinations; destroying evidence and falsifying records; employing ex-Nazi research scientists; having connections with the Mafia and organized crime networks and supplying funds, arms and illegal drugs to foreign combatants in support of their fighting proxy wars with adversaries who pose potential national threats.

Of organizations such as the United States Department of Agriculture (USDA), the Food and Drug Administration(FDA), the Drug Enforcement Agency (DEA) and the Bureau of Alcohol, Tobacco, Firearms and Explosives (ATF) that protect the nutrition, health and safety of human beings through the regulation and oversight of the food supply, medications, drugs and controlled substances and the enforcement of violations of the laws and regulations governing their approval, production, safety, efficacy, distribution, trafficking, sale, recall and discontinuance.

Of the mysterious heart attacks, 'accidents', fallings out of high windows, suicides, deaths and disappearances; abductions, expulsions and exiles; captive-takings and imprisonments; executions, assassinations and dismemberments that continue to occur; especially of those targeted outspoken, dissenting, whistle-blowing and potentially testifying human beings; those advocating democracy and equality and championing human and civil rights in autocratically- ruled totalitarian nations and countries.

Of those targeted human beings who are revolutionary and opposing, critical and exposing of, or in some way posing perceived threats to, the governmental power and control structures, media propagandizing and censorship and the dissimulation and cover-ups and clandestine, corrupt, conspiratorial absolute rule, control and oppression of fascistic plutocrats, oligarchs, tyrants, despots, patriarchs, autocrats and dictators.

Of the various sanctions imposed upon autocratically- and dictatorially-led nations by the international community that do not appear to make any real or significant difference in the 'business as usual' of such nations; while the human beings residing in them continue to suffer ongoing abusive injustices, oppressive controls and suppressed freedoms that their occasional courageous protests, rallies and mass demonstrations often fail to influence, change, correct or end and result in mass arrests, detainments and incarcerations.

Of some autocrats and dictators who have resigned under pressure or have been deposed and exiled and their governments overturned by popular democratic vote; political juntas and military coups; strong collective opposition and demonstrated social protest; open rebellion and revolution and serious credible death threats to them and their family members.

Of the civilian, governmental, medical, military and police human rights groups, aid workers and peace keepers who have gone to various needed places in the world and who have been killed, attacked, imprisoned, driven away and forced to evacuate in the process of assisting human beings suffering from the ongoing destructive and traumatic effects of war; humanitarian crises; epidemic disease and lack of medical care; tyrannical, autocratic and dictatorial rule and governmental factionalism; poverty, malnutrition and starvation, etc..

Of the mature, wise, rational and compassionate world leaders, national and international foreign ambassadors, ministers and diplomats who value the lives, health, safety, security, success, freedom, peace and happiness of the human beings who inhabit their countries and nations and who exercise restraint with regard to inducements to act in conflicts with military interventions that provoke and escalate into wars and who instead use diplomatic negotiations that result in de-escalations, settlements, resolutions, compromises and ultimate peace.

Of my wondering about how many precious human beings, blessed human souls and sacred human lives need to be uprooted and displaced; destroyed, taken, lost and become 'collateral damage'; injured, wounded and crippled and ruined, dispirited and desecrated in wars before there is a significantly influential global shift and transformational turning around of our ordinary ego-conditioned human consciousness to a more collectively awakened, enlightened, wise, inclusive, egalitarian, compassionate and peaceful one.

Of my wondering how many 'wake-up calls' need to occur before leaders, governments and military organizations, 1) actually do 'wake up' from the dreams, nightmares, illusions and delusions of what real and true human being and living are about and 2) humanely devote their attention, resources, strengths and efforts to world health and peace; global co-existence and cooperative, collaborative and synergistic relationships between countries, nations and their citizens.

Of my hope for some consequent significant measure of a corrective reduction of, and significant liberation from, offensive wars (not 'defensive' wars necessary for protection from aggressors, attackers, invaders and occupiers) initiated, justified, rationalized and perpetuated by power hungry leaders, foreign policy makers and war-mongering jingoists.

Of the need for heightened planetary environmental consciousness and awareness about biodiversity and ecological interdependence and balance and for curtailing commercial activities that are destroying virgin lands, rain forests, ecosystems, natural habitats and the homelands of first nation and indigenous peoples along with their cultural tradition, creation stories, memory, knowledge, wisdom, plant medicines and healing practices and that are resulting in increasing species endangerment, degradation and loss; including our own human one!

Of the unfortunate reality that the lives of a growing number of human beings, especially indigenous peoples; living in an increasing number of countries throughout our world; are not going well, securely, successfully, optimally, freely, peacefully, contentedly and happily. Some of which is due to the effects of deforestation and land-, water- and power-grabs; corporate and investor greed, privatization, profiteering and gentrification; hierarchical power differentials and internal conflicts; ideological, political and civil wars and economic disparities, inequalities and inflation in the countries in which they are living.

Of the consequent result that an increasing number of human beings are suffering and struggling with imposed individual and mass uprooting and displacement, homelessness, deprivation, impoverishment, malnutrition and starvation and end up living undignified, unprivileged and uncivilized sub-human and inhumane lives that are compelling them to resort to migrating, protesting, striking, rioting, violence and criminal activity such as vandalism, home invasions, store and auto break-ins, shoplifting, smash-and-grab and flash-mob robbery, looting and carjacking; scamming, mail and identity theft, ransomware attacks, etc. in order to survive.

Of the safeguarding, preserving, maintaining and sustaining and the respectful regard for and intimate sharing of the validity,

dignity, integrity, value, worth, decency, civility and humanity of our universal inborn True Tao-Nature and our blessed
endowment and precious treasure of sacred human incarnation
and human life.

Of the possibility, reality and actuality of our collective
human experiencing of the light, freedom, peacefulness, intimacy
and happiness:

1) that is an ultimate enlightened awakening after far too
many 'wake-up calls';

2) that follows from the effective resolution of conflict and
the successful ending of war and

3) that each and every one of us innately, rightfully, clearly,
deeply and fully deserves by virtue of being a Human Being living on Planet Earth in harmony with each other and with the
bountiful gifts of Nature and perfectly heart and soul-centered
between the solid supportive groundedness of Earth and the vast
encompassing spaciousness of Heaven.

Of those of us human beings who are still able;

1) to clearly, deeply and fully safeguard, cherish, treasure
and nourish;

2) to purely, simply and wholeheartedly create, enjoy and
fulfill and

3) to sheerly, utterly and intimately share the awesome,
amazing and astonishing mystery, miracles, marvels, magnificence, magic and majesty of our exquisitely human being and
perfectly evolving human living in spite of everything that is
and/or is not currently happening within and between ourselves
and in our lives, our world and on our planet.

And of, finally and most critically, our dualistic abstracting,
objectifying, construing, labeling, separating, dividing, alienating ('other'-making), projecting, judging and blaming ego-
minds and ego-selves that make either-or and mutually exclusive
discriminations, comparisons, contrasts, oppositions, opinions,
preferences, conclusions, etc.;

1) that are the root source and principal cause of differences, disagreements, arguments, disputes, contentions and antagonisms between human beings, groups, organizations, societies, countries, nations, governments and leaders and

2) that involve systemic inequalities, seemingly intractable issues, failed democratic and diplomatic negotiations, deep-seated animosities and hostilities and apparently long-standing unresolvable conflicts that result in ongoing adversary- and enemy-making and further escalating into armed military conflicts and full-blown wars.

> A sympathetic recognition
> of the lamentable tradition,
> of the conflicting friction,
> of the warring addiction,
> of the human condition.

> An unsympathetic recognition
> of the dictators' ambition
> and warring conviction,
> of the warrors' attrition
> without leaders' contrition.

YOUR REFLECTIONS

The definitions and etymologies of the following Chinese language characters may assist you, the reader, in understanding and appreciating the spirit in which an opportunity and space are respectfully and humbly offered and provided to you for your periodic reflections as you read through this book as extraordinary and intimate human beings of heart.

NI/NI – You/*Your*

你

人 Jen/Ren – *human being*
+
小 Hsiao/Xiao - *small*

Typically used for common human beings of ordinary social status. (Here, Hsiao connotes a lower status).

NIN/NIN - You/Thou/Your/*Thy*

您

心 Hsin/Xi – *heart*, center
+
你 Ni/Ni – *you*

Honorifically and respectfully, deferentially and politely used for elders, superiors, intimates and extraordinary human beings. (Here, Hsiao connotes humility).

CHAO/ZHAO – *To reflect*, shine on, light up, illuminate, enlighten

照

昭 Chao/Zhao – *bright, shine forth*
+
灬 Huo/Huo – *fire*
or
日 Jhi/Ri – *sun*
+
召 Chao/Zhao – *summon*, call forth
刀 Dao/Tao – *knife*

口 K'ou/Kou – *mouth*

(Originally, Chao meant criticism but now connotes sharpness, keenness and incisiveness).

Your reflections are the summoning and calling forth of the sunlight and firelight that is shining upon and is illuminating and enlightening your True Nature and your whole being.

May your reflections mirror your utmost, innermost, centermost and uppermost identity as an extraordinary, real and true human being of heart.

YOUR REFLECTIONS

Prologue

War, The Army and the Military

As a young boy during the time of World War II, I was disquieted by the awareness of wars going on; news about wounded, dying and dead soldiers and pictures of invaded countries and destroyed cities. I had three close uncles in the armed forces, one in the U.S. Navy, one in the U.S. Army, one in the U.S. Air Force. At several visits, they each shared scarey, poignant and disheartening stories about their first-hand experiences of being in the war.

Uncle Joe was on a destroyer escort ship that was required to head off any enemy torpedoes directed at destroyer ships. The ship's cook reportedly placed raisins on top of bugs when making bread. Uncle Johnny manned an LST, landed on a beachfront, opened the front of the craft only to see exiting soldiers immediately gunned down by the enemy. Uncle Vince was a bomber pilot who had his co-pilot blasted to pieces right next to him by enemy anti-aircraft fire.

In our small and ordinarily quiet Chicago area suburban village, there were frequent air raid drills. Sirens started blaring. The town streetlights were all turned off and I remember seeing searchlights reaching upward, sweeping around scanning the night sky. We were mandated to turn off our home lights and to shelter in place. Our family huddled together under the kitchen table until the ending sirens sounded.

I remember that my father escaped being drafted into the military because his steel manufacturing company was involved in essential war efforts. I remember my mother being a gray uniformed Red Cross volunteer. I remember peeling labels off of tin cans and stomping them on our kitchen floor to be recycled for use in the war effort.

Later on, I remember my older brother being drafted and serving in the U.S. Army infantry when our country was between wars. While not being involved in combat anywhere overseas, he shared stories about basic training cruelties, brutalities, suicides, soldiers fighting with each other and going AWOL and a soldier crawling on his back next to him and having his knee caps shot off by machine gun errors during a training exercise.

I was drafted into the U.S. Army while being a Ph.D. Clinical Psychology graduate student which didn't qualify me for a deferment. My induction physical was a sham. Much of my 'physical examination' consisted of the self-reporting of my health status. Vital signs data were obviously fudged and the chest x-ray machine had no film in it. I was found to have 'pes planus 3rd degree' aka 'flat feet', which was noted but ignored as far as induction went.

Primarily to avoid being a regular infantry soldier and going through basic training, I signed up to be a commissioned officer in the U.S. Army Reserve Medical Service Corps as a clinical psychologist. While serving at Letterman Army Hospital, I observed the attitudes and behavior of numerous military personnel of various ranks and experienced the vicissitudes of what I came to regard as a 'military mentality'.

It was not during wartime and many of the male personnel were dissatisfied with 'paper shuffling', feeling emasculated, recounting 'war stories' and regretting that they were not currently actively engaged in fighting an enemy somewhere.

I sat in on a court-martial proceeding. It was a mock hearing and the soldier was asked to leave the room and to wait outside until called back in to receive his ultimatum. The committee chairperson had already decided ahead of time to dishonorably discharge the soldier and told us committee members to sit around for awhile to give the impression that we were deliberating the case.

When I asked the Colonel Chief of the Neurology Service why some adolescent children of military families were being lobotomized to 'treat' behavioral disorders, he replied, 'We may not know what we're doing, but we're never unsure of ourselves'.

Throughout his blackboard presentation, a senior officer/psychiatrist repeatedly misspelled the words 'psychiatry', 'psychology' and 'psychotherapy' (consistently omitting the letter 'h'). Of course, none of the snickering junior officers/psychiatric residents and psychology interns had either the insensitivity, courage or stupidity to point it out.

When I had reason to knock on closed staff office doors to do clinical business, I typically heard desk drawers being hurriedly slammed shut before being given the 'come on in'. (In those days there were no computers used for 'pornographic' distractions and obsessions and 'girlie' magazines were relied upon instead).

I enthusiastically and thoroughly enjoyed my psychodiagnostic and psychotherapy work with children, adolescents, adults, couples and families of military personnel and was highly productive. At one point; I was told that I was 'seeing too many patients' and 'showing up' my peers and to 'cut back' on my case load.

I had an 'authority problem' with the Colonel Chief of the Clinical Psychology Service on a number of occasions and did some 'acting out'. On one such occasion, he was discussing that during WWII the ink on batches of Rorschach projective testing cards was different which affected consistent diagnostic assessments and determinations. I blurted out, 'War is hell!'.

On another occasion, the Chief advised me to make a clinical treatment decision based upon military protocol rather than clinical efficacy which I refused to do. I said, 'If I don't do it are you going to court martial me for insubordination?' and left his office, noticing his dumbfounded facial expression out of the corner of my eye on my way out the door.

By and large the leadership and staff of the Neurology, Psychiatry, Child Psychiatry and Clinical Psychology Services were competent medical physicians, neurologists, psychiatrists, psychologists and social workers who were more professionally-oriented rather than military-oriented and who had chosen serving in the military mainly for the provided advantages, numerous benefits, potentially desirable assignments and the ability to retire at full pay while still being relatively young.

I decided to leave the military after serving my five year commitment and was honorably discharged as a Captain in the U.S. Army Reserve Medical Service Corps.

On Harming, Killing and Guns

As a boy, I enjoyed the benefits of being in Nature close to our home and interacting with the many creatures down at the Desplaines River that ran through our town; oddly enough named Riverside, Illinois. I caught various insects, butterflies, moths, caterpillars, spiders, dragonflies, bees, wasps, beetles, bugs, snakes, crabs, frogs, fishes and turtles; played with them for awhile and then gently released them back to their native habitats. I never harmed or killed any of them. (I even used doughballs for fish bait rather than dug-up worms).

A high point in my activities was housing a plump caterpillar in a soil and twig-filled wooden cigar box and observing its transformations into a spun white cocoon, shiny wiggling chrysalis and beautiful emerging yellow and black tiger swallowtail butterfly. A low point in my activities was inadvertently having a mud turtle and two crabs share a water-filled basin one day and waking up the next morning to find four crab claws and two empty crab shells floating in the water.

I did swat flies and mosquitoes and did watch my parents killing household flies, moths, spiders, ants, earwigs, silverfish, etc. and spraying trees, shrubs, plants, flowers and garden insects with toxic pesticide and fungicide chemicals. Once, my father unearthed a nest of garter snakes and aggressively hacked them all to pieces with the sharp edge of his shovel.

As a boy, I had squirtguns and double-holstered cap pistols and detonated tin cans with 'cherry bomb' and 'two-incher' fire-crackers. I had knives that were used for playing mumbly-peg, throwing at trees and telephone poles and whittling wood and carving bars of soap.

I had a B-B gun and, one time, did shoot at and kill a robin, which I deeply regretted. A ricochet from my B-B gun accidentally hit a neighbor girl in the corner of her eye. I was shot at by a 'friend' using his B-B gun. I got rid of the gun and vowed never to shoot one near, or kill, another living thing. This was well before knowing about and making the Hindu-Buddhist vow of 'ahimsa' or non-harming/killing.

When in college, I did purchase a slide-action Winchester .22 caliber rifle and a lever-action Winchester Model 73 .30-.30 saddle carbine, mainly for their aesthetics, bluing and American black walnut stocks. Once, I and a fraternity brother went to the city dump at night, with flashlights tied to our .22 caliber rifles, with the intent of hunting rats but none were injured or killed. I only fired the .30-.30 rifle one time at a shooting range.

When in U.S. Army, I was given an accurized Colt .45 caliber pistol by a fellow officer in trade for my winter military coat. Later on, I did purchase a Berretta .32 caliber pistol, again mostly for the aesthetics and wooden grips. I sold the Colt to a gun shop, never fired the Berretta and it and the Winchester .22 rifle were recently purchased by a dear friend. The Winchester .30-.30 was recently sold to a gun shop along with a B-B gun that I had purchased for my youngest daughter (when she turned twelve!) for target shooting. It feels good not to own any firearms.

CONFLICT AND FIGHTING

During my early years, I was deeply disturbed by wars and warring and confused by what appeared to be a permissible paradoxical 'ethic' of 'killing for peace'. I was against capital punishment, putting human beings to death and 'killing killers', rather than incarcerating them for life without the possibility of parole.

When I first learned about justifiable wars and '*Lex Talionis*'; the 'Talion Law or Principle' of reciprocal justice, retribution and punishment that were equal to but not greater in kind than the aggression and offense, i.e., 'an eye for an eye'; I had difficulty accepting the reality that this often was not in fact the case.

I never saw or heard my parents arguing. On one occasion, my 8-year older than me brother did get into a physical fight with my father before leaving home to be on his own. I never had any arguments, conflicts or real fights with my brother, except over second-helpings of food, who could first make preemptive strikes and raids on the refrigerator and who got to be first starting a board-game.

One time, I remember my mother stomping on my toy train out of her frustration with me. One time I said something to my father that made him order me to my upstairs bedroom. He booted me in the butt as I was climbing up the stairs. After I was given permission to come back down, I told him that he should not have done that if he loved me. He then went up to his own bedroom and I later found him there sitting on a chair and weeping.

In elementary and grade school, I was threatened and bullied by two fellow students and avoided encounters with them by taking a circuitous way back home from school. I was frequently punched on the upper arm by 'friends' adhering to the favorite 'like it or lump it' dictum. I remember one winter after school witnessing a fist fight between two fellow students and watching the red blood dripping from their noses onto the pure white snow beneath their feet.

Once, during a mudball fight with some neighbor boys, I was hit in one eye and narrowly missed losing my vision. On one occasion in high school, I said something to a classmate during a ping-pong game that made him come over to me and try to hit me. I held down both of his arms rather than fight with him until he cooled off.

In college, I argued with one of my fraternity brothers on a semi-regular basis. On night while being driven through the college town, I yelled something out of the car window that was misheard by a boy who then came over to the car and punched me in the face, chipping a front tooth.

As a youth, adolescent and 'adult', I was not assertive, aggressive or assaultive and did not initiate or directly engage in any physical fights. I typically either successfully stopped fights and conflicts or avoided them completely.

In relationships with my three daughters, I had occasional differences and disagreements with them but never fought with them. I attempted to mediate their own conflicts between them and with peers and to facilitate peaceful resolutions. I am still estranged from my oldest daughter following her expressing extremely negative and hateful feelings towards me and making vitriolic and damning criticisms of me which I have chosen not to accept, respond to, work out or defend against.

I avoided conflicts and did not share conflictual feelings with my two ex-wives and with the several women with whom I lived concerning their either not wanting to have children, already having children or not wanting to have any more children and instead distanced myself from them and forewent developing an intimate connection. This, along with my not identifying with many expected traditional social male roles and the creative engineering of my strong unmatched inner feminine anima, ultimately resulted in mutually unsatisfactory and dissatisfying relationships, 'irreconcilable differences' and eventual divorces and separations.

On a few regrettable occasions; as demonstratively modeled by some of my Sicilian-Italian relatives; I did lose my temper, raise my voice and pound my fist on a table when frustrated, angry and upset; which behavior undoubtedly caused my partners to become fearful, created more distance between us and also significantly contributed to the eventual divorces and separations.

Overall, I do not regret the ways in which I lived my past

relationships but do occasionally wonder if I would have personally developed in more mature and complete ways as a human being had I not avoided conflicts and had perhaps instead confronted them directly and understood and experienced their value in contributing to more real, whole, intimate, fulfilling and sustained relationships.

GAMES AND ATHLETICS

As a youth, the dicegames, cardgames and boardgames that I played were ones that I simply enjoyed playing and that involved either chance, the spin of a dial, the luck of the draw or my own ability *vis-à-vis* myself rather than being in competition with, defeating an opponent and winning a game, e.g., Yahtzee, 'War', Old Maid, Scrabble, Monopoly and Checkers. I was not interested in poker, gin rummy, canasta and chess because the calculating activities and bluffing strategies involved and required to win in 'gamesmanship' detracted from enjoying playing. I never cheated in playing games nor gloated over and flaunted winning them.

In athletic activities and sports, I played and enjoyed darts, archery and target shooting, skating and bicycling, track and field, golfing and bowling, baseball, basketball, volleyball, tennis, badminton, ping-pong, pool and billiards. Darts, archery and target shooting were about accurate aiming and hitting the bulls-eye and track and field was about ability, skill and endurance.

Skating and bicycling were about balancing and maneuvering. Golfing and bowling were about accurate aiming and sinking putts and making strikes. Baseball and basketball were about accurately pitching strikeballs and making set shots. Volleyball, tennis, badminton and ping-pong were about accurately placing shots. Pool and billiards were about seeing angles, accurate aiming and correct 'touch'.

Football was about trying to be comfortable sitting on hard benches for most of the game. Driving a car was about shifting gears and braking smoothly and timely, staying in the middle of lanes and parking in the middle of spaces. Dancing was about free-form moving to rhythms rather than executing the specific forms and movements of particular dances.

In all of the activities, my only 'competitive spirit' was not in winning over others but was essentially in relation to my own ability, skills, performance and proficiency most all of which primarily involved taking accurate aim and making precise and coordinated movements that would result in successful outcomes.

PHILOSOPHY, PSYCHOLOGY AND PSYCHOTHERAPY

At the age of seventeen, as a college freshman and psychology major, when I read my first copy of Lao Tzu's *Tao Te Ching* purchased on day one at the bookstore; I was pleased to note the verses that professed anti-war values, a restraining and refraining conflict ethic, warfare as an unavoidable last resort and concomitant stances of reticence and hesitation and feelings of sadness and regret over engaging in warfare even when being victorious.

A similar morality ran through Chuang Tzu's Taoist tales and Lieh Tzu's Taoist stories regarding the ill-fate of autocratic and militant leaders, internecine warring feudal states and aggressive and invasive warlike and warring human conduct. My life-long personal and career interest in the teachings and practices of the ancient spiritual tradition of Chinese Taoism and of Buddhism, 'Eastern' Philosophy in general and Phenomenology and Existential Philosophy in particular began at that time.

In studying, learning, teaching and practicing psychology and psychotherapy, I was thoroughly competent and proficient in psychopathology, psychodiagnosis, behavioral observations, interviewing, case formulations, treatment planning, methods and techniques, intervention strategies and outcome evaluations.

But I always saw named psychological conditions as only one

particular way, perspective and level of observation and way of relating to human beings and never objectified or 'pathologized' them in terms of abstract psychological theories or manipulated them *via* concrete psychotherapy techniques.

There is an old joke about human beings coming to a psychotherapy appointment and suffering the realities of having behavior analytically interpreted instead of simply accepted as being what it is without a psychodynamic meaning. If you are early, you are anxious. If you are late, you are resistant. If you are on time, you are compulsive. No way to 'win'!

I maintained a healthy skepticism about the need for, and the efficacy of, nosological categories and psychodiagnostic labels, which were most often only used for purposes of making insurance claims but stayed with human beings for their whole lives. What was life like before the institution of diagnostic categories for 'disorders'?, e.g., human beings who are irritable in hot summer days and sad in cold winter ones now have a 'seasonal affective disorder'.

I was disappointed by the proliferation of more and more 'conditions' being created and labeled, 'spectrums' being designated, e.g., autism and ADHD, and the bar continually being lowered for those manifesting their symptoms and being assessed to 'qualify' to be so identified. Now, most everyone has some kind, degree and level of anxiety or depression. Will more and more human beings soon have BBD and LLD, a Being Born Disorder and a Living Life Disorder?

When training in and teaching the approaches to, and methods of, psychotherapy; I preferred ways that were more phenomenological, conscious awareness- and emotion-focused, experiential, existential-humanistic and transpersonal rather than ones that were more psychoanalytic, cognitive and behavioral and which I regarded as too abstract, theoretical and manipulative. The approaches of some trainers and practitioners were even called 'strategic' and involved various *in situ* and often painful 'aversive deconditioning', 'prescribing symptoms', 'over-doing' and 'double-binding'.

In the psychotherapy/counseling attending relationship process, I focused more on facilitating self-reflecting, self-aware-ness, self-understanding, self-accepting and self-actualizing and conducting and orchestrating the relationship rather than con-trolling, directing and manipulating it.

I eschewed military metaphors for considering how to deal with physical conditions and psychological issues, e.g., fight, battle, defeat, beat, conquer, destroy, kill the illness, disease, condition, symptom, trait, habit as an 'enemy'.

I preferred methods such as visualizing and befriending con-ditions, issues and symptoms and discovering and understand-ing the purpose that they serve for maintaining the integrity, stability and growth of the whole human being; e.g., by letting them 'speak' and consulting, interacting, communicating, dia-loguing and role-playing with them.

Throughout the years; in clinical psychological, psycho-therapeutic and counseling practice; I assisted many children, adolescents, adults, couples and families variously deal with dif-ferences, dissension, disagreements, disputes, arguments, quar-rels, fights, conflicts, frustration, anger, rage, aggression, abuse, violence, hatred, hostilities and suicidal and homicidal thoughts and impulses and some acting out.

I generally advocated for mutual understanding and empa-thy and 'taking the high road' of egalitarian, collaborative, com-promising and peaceful conflict resolution and reconciliation. I supported non-adversarial arbitration in lawsuits involving con-flicting parties and favored bi-lateral diplomatic negotiations in resolving national and international conflicts. I also participated in numerous peaceful anti-war civil protest demonstrations and marches.

I did not regard private counseling practice as a 'business', set fees on a sliding scale basis, wished we could barter and was uncomfortable monetizing and commercializing psychotherapy and earning money by capitalizing on human being's problems, even though 'services rendered' were critical, necessary and obviously effective, beneficial, meaningful and worthwhile.

I was aware of how much of 'business' in general involves a parasitic and predatory relationship between human beings. I typically reframed psychological crises more as spiritual awakenings and, as quickly as possible, assisted 'patients' and 'clients' in becoming independent of fee-for-service psychotherapy and in finding helpful strengths within themselves and appropriate human resources in their lives.

I was always aware of, and sensitive to, the power differential inherent in psychotherapeutic relationships and quickly negotiated and resolved the occasional power struggles occurring as a result of the role inequality and of transference projections. It seemed that some of the psychotherapy 'issues' were generated by the inherently separate, disparate, unequal and hierarchical structure of the very relationship itself. The 'therapist' appears to see and to know more about the 'patients' and about their being a healthy self and living a fulfilling life than the 'patients' do.

The reasons that unmandated human beings initially sought out and participated in some kind of 'therapy' seemed to be due: 1) to anxieties about, depression over and fears of; danger, isolation, low self-worth, insecurity, inadequacy, failure, illness, pain, injury and dying, 2) to the unquestioned belief in the unassailable authority and salvational power of so-identified 'professionals' (often wearing long white coats), 3) to an ignorance of the unique and synchronistic developmental meaning and the growth potential and opportunity of concerns, symptoms and conditions and 4) to an ignorance of human beings' innate and inherent power and ability to consciously and fully understand them and to heal themselves.

The treatment models that I valued and used were those of prevention, early and crisis intervention, harm reduction and damage control. With needing to find competitively more clever ways of connecting with referral sources in order to obtain clientele, the advent of managed health care, the requirement of stringent initial disclosures and certain procedural rules, a general increase in legal issues and malpractice law suits and fees exceeding one-hundred dollars per fifty-minute hour; doing psychotherapeutic counseling was no longer a very enjoyable activity and I gradually retired from private practice.

For me, the profound gift and deep significance of being an available, approachable and accessible fellow human being and of effectively and efficiently making a valuable, meaningful, worthwhile and beneficial difference to fellow human beings as a somewhat wise and certainly compassionate spiritual friend and fellow Wayfarer could be more happily enjoyed and comfortably lived through some freer and more natural, satisfying, gratifying and fulfilling ways in the world.

TAOISM AND *THE ART OF WAR*

During immersion in the life-changing, self-transforming and spiritually evolving Taoist philosophical and spiritual tradition, literature, wisdom and T'ai-Chi and Ch'i Kung practices; I inevitably became aware of Sun Tzu's *The Art of War* but had never read it, even though my primary interests in ancient Chinese Taoism are from the Spring and Autumn (770-475 BCE) and the Warring States (475-221 BCE) Periods, the times when Lao Tzu, Sun Tzu, Chuang Tzu and Lieh Tzu lived and wrote.

I had disregarded *The Art of War* because I thought that it was about warfare as such, which I was against and in which I was profoundly disinterested. I was pleased to later discover that, quite to the contrary, its essential teaching relates to successfully defeating opponents, resolving conflicts and winning wars through knowledge, strategic plans, tactical maneuvers and intelligence-gathering rather than by openly confronting, forcefully engaging and directly fighting with them.

MYSELF AND WAR

It has taken me what now seems like a long time to modify, integrate and balance my naturally idealistic and optimistic predispositions with ones that are more realistic. This has initially involved the reluctant acknowledgment, and now the compassionate appreciation, that disagreement, opposition, antagonism, conflict and war are an unfortunate part of human nature and the human condition and that subduing, overcoming, defeating, conquering, vanquishing, destroying and killing so-identified 'enemies' are tragic activities of human beings and the human condition.

A case in point: A part of Passage 69 of Lao Tzu's *Tao Te Ching*, is usually translated as 'There is no greater calamity than underestimating enemies and losing our treasures (unconditioned love, conserved resources and restrained precedence)'. In my own translation, I took the liberty to 'translate' this sentence as 'For us, there is no greater calamity than creating and fighting 'enemies' and losing our treasures'. Doing so reflected my feeling that even seeing, identifying and naming fellow human beings as 'enemies' in the first place was already the calamitous loss of unconditioned love, energy resources and humility.

It has taken a long time to accept the unfortunate and disheartening reality that, and the extent to which, some of us human beings may inevitably and continually conflict with and violently oppose; take advantage of and exploit; deceive and betray, violate and abuse and injure and kill others of us human beings. And that the latter need to be protected; to learn how to effectively strengthen and defend themselves and how to courageously extricate themselves from toxic superior-inferior, leader-follower, dominating-subjugating, controller-controlled, manager-managed, master-slave, host-parasite and predator-prey kinds of alienating, dehumanizing and disempowering dualistic, oppositional, oppressive and conflicted relationships.

Currently, near the completion of my own life's Wayfaring journey and sojourning pilgrimage, I am disinterested in engaging in unproductive disagreements and disputes, meaningless arguments and conflicts and senseless battles and fights that ultimately achieve and accomplish little or nothing that is real and desecrate and destroy much that is true.

I am more involved in living Taoist principles: 1) of the Ultimate Reality of Tao and the power of inborn True/Tao-Nature and its Virtuosity/Te, e.g., being a Spirit-body or human Soul, 2) of Wu-Wei Ch'i energy kinetics of not-forcing actions and flowing with seamless, frictionless and effortless activity, e.g., like riverstreams do and 3) of Yin-Yang Ch'i energy dynamics of complementary rather than opposing and conflicted relationships between alternating bipolar unities, e.g., love-hate, harmony-conflict, peace-war.

In this regard, I thought that it is 'high time' that I explore the yang-side of the matters of opposition and conflict and seriously consider the early, classic, well-respected and widely-employed ancient Chinese classic treatise on the Tao/Way/Method of warfare, namely *The Art of War* by Sun Tzu, the renowned military general, philosopher and strategist.

I consulted the translated writings listed in the References section of this book and decided to write a rendition of my own. I consulted bi-lingual editions and looked up the meanings and etymologies of a number of key Chinese language characters of the original text and discovered, as is frequently the case, that translations were interpretative and often very different.

I decided to add commentaries to the rendition of the text that are psychological and sociological in nature and characterize both intrapersonal, interpersonal and non-personal oppositions and conflict. I therefore titled and subtitled the book Sun Tzu *The Art of War: Psychosocial Commentaries. A Wayfaring Sojourner's Rendering of the Way of Warfaring.*

I AND THEE

Writing this book has been an engrossing, enjoyable, illuminating, significant, important and meaningful integral experience of wisely, harmoniously and compassionately remembering and bringing together, reviewing and reflecting upon, uniting and balancing and transforming and partially transcending my own past and current history with issues of conflict and successful and unsuccessful ways of avoiding, dealing with and/or resolving them.

I was pleasantly surprised to experience how natural and easy it was to feel familiar with and to understand the general principles, gist, sense and 'drift' of Sun Tzu's *The Art of War* after historically having had such strong objections and aversion to war and such negative opinions and feelings about war.

I better understand, compassionately accept and appreciate the seemingly undeniable, inevitable and unavoidable human realities of both conflict and resolution and war and peace within myself/ourselves, between myself/ourselves and fellow human beings, in our society and between nations of the world.

I better understand, compassionately accept and appreciate the ways in which my past avoidance of head-on, toe-to-toe, in your face confrontational conflict has significantly positively and negatively impacted, influenced, affected, contributed to and determined my sense of self and intimacy and my overall personal growth, professional development and spiritual evolution.

As human beings; among many other factors involved, choices and decisions made and unmade and actions taken and untaken; we are products both of that which we accept, acknowledge, affirm, confront and allow and of that which we reject, deny, negate, avoid and refuse within ourselves, within our lives and within and between our fellow human beings.

Everything considered; I appreciate, enjoy, value, treasure and cherish ending up being:

1) a basically interested, attentive, caring, kind, tender, gentle, soulful and spiritual human being with a well-developed and integrated inner feminine/anima nature;

2) a biologically-gendered male human being who is relatively disidentified and free from a traditionally alpha-male personal identity and many conventionally masculine social roles;

3) a caring and empathic human being who currently still sadly regrets and compassionately grieves the traumatic injuries to healthy human beings, the wasteful losses of precious human lives and the complete destruction of created architecture in armed conflicts, military combats and wars and

4) an essentially peaceful and relatively free and happy human being who has experienced the value, advantages and benefits of knowing when and how to creatively, artfully, skillfully, effectively; and at times strategically and tactically; move below, behind, between, around and above obstacles, obstructions, disagreements and conflicts that are in my Way, as well as when and how to confront them openly, deal with them concretely and directly and work through them completely.

I hope that you, the interested readers of this book:

1) will awaken to, remember, resonate with and reflect upon and better accept, understand and integrate some significant personal and interpersonal experiences and meanings of your own with respect to matters of conflict and war and

2) will discover and appreciate some new, enlightening and enlivening awarenesses concerning integrating peaceful Wayfaring and conflicted warfaring within your precious selves, lives and relationships with fellow human beings.

Grace, blessings, good fortune and gratitude and peacefulness, freedom, intimacy, happiness and joyfulness as your human birthright; as your most valid, dignified, worthwhile and deserving Self and True Tao-Nature; in your most sacred, precious, treasured and cherished life and throughout your most unique, satisfying, gratifying and fulfilling Wayfaring journey.

Thank you for the opportunity to share with you from the wholeness of Being, the centerness of Self, the openness of Heart, the deepness of Soul and the spaciousness of Spirit. For you with love! Please enjoy!

Raymond Bart Vespe
Scotts Valley, California
Autumnal Equinox, 2024
UN International Day of Peace

It is possible to live
without war and conflict.
It is possible to live
in peace and harmony.

There is no real 'enemy'
to war against.
There is no true 'other'
to conflict with.

There is no real 'this' *or* 'that'.
There is no true 'us' *or* 'them'.
We all essentially are
One and the same human beings,
the same and One humanity.

And we can choose
Either the love of power
Or the power of love.

Your Reflections

Sun Wu Tzu
The Art of War
Chinese Characters

孫

Sun/Sun
Grandson
Tzu/Zi – Child/Boy/Son +
Yao/Yao – Small/Tender +
Hsiao/Xiao – Small/Young

武

Wu/Wu
Military/Martial
Chi/Zhi – Halt/Cease +
Ko/Ge – Spear/Lance

子

Tzu/Zi
Child/Boy/Son
Seed/Sir/Philosopher/
Master/Gentleman/
Young/Tender

兵

Ping/Bing
War/Military/
Soldier/Weapons
Pa/Ba – Eight +
Chin/Jin – Ax/Battle Ax

法

Fa/Fa
Way/An Art/Method/Law/Plan
Shui/Shui – Water/Fluid +
Ch'u/Qu – Go
(Ancient characters for Fa/Fa =
Chi/Ji – Together + Cheng/Zheng – True/Upright)

Introduction

Sun Tzu's *The Art of War* is the oldest military treatise and classic of strategic and tactical warfare. For over two-thousand five-hundred years; it has been studied, consulted and implemented world-wide by leaders of countries during times of national and international conflicts and wars. Recently, it has been applied, e.g., in martial arts, healing arts, business activities, athletic competitions, legal proceedings and self-development. The current application is to intrapersonal and interpersonal psychological and psychosocial conflicts.

Authorship

The Art of War, Ping/Bing Fa, is authored by Sun Wu Tzu/Zi, Sun the Martial, Master Sun, who reportedly is born and lives sometime during the mid to late 6th and early 5th Centuries BCE (c. 544-496 BCE) in the Chinese State of Ch'i/Qi. He reportedly is an outstanding army commanding general, philosopher and military strategist who never lost any of the wars which he and his armies either strategically avoided or were successfully engaged in. He may have been a military advisor to the King of the Chinese Wu State toward the end of the 6th Century BCE.

Little is known about the life of Sun Tzu/Zi other than some honorific references relating to various battles and to *The Art of War* being consulted by military leaders. References to him are absent from comprehensive historical records of his supposed lifetime.

There are controversies among some sinologists as to whether Sun Tzu/Zi is an actual person, a generic designated title given to outstanding military leaders or is a name adopted by and/or conflated with an actual familial-related Sun Pin/Bin (c. 380-320 BCE), who also authors a text entitled *The Art of War*.

THE TEXT[1]

The Art of War, originally titled *Master Sun's Military Methods*, is controversially reported as being written sometime between the late 6th and early 5th Centuries BCE (c. 515 BCE), i.e., either sometime in the late Spring and Autumn Period (770-475 BCE) or sometime during the early Warring States Period (475-221 BCE) of Chinese history. The latter would closely postdate Lao Tzu (c. 6th Century BCE), coincide with Confucius (551-479 BCE) and predate Chuang Tzu (365-286 BCE) and Lieh Tzu (c. 4th Century BCE).

However, the philosophical foundation, framework and principles of *The Art of War* are decidedly more Taoist rather than Confucian. Tao/The Way and its virtuosity/Te are acknowledged as method, power and skill; Yin-Yang complementary bipolar-unities are dynamically balanced and Wu-Wei flexibility and effortlessness are kinetically enacted.

There are disputes among some sinologists (as is the case with Lao Tzu and the *Tao Te Ching*), as to the exact dating of the text and, in this case, whether one person named Sun Tzu/Zi is the sole author of *The Art of War* or it is: 1) a later compilation made by the Sun Clan and his purported grandson Sun Pin/Bin or 2) composed of additions to an original text made by other individuals and accreted over a period of time (as is the case with the Outer and Miscellaneous Chapters of *The Chuang-Tzu*).

Using historical references as to the time frames of: 1) the formation of a standing army of trained mass infantry; the reasons why, the places where and ways in which wars were fought; change in military leadership personnel; soldier conscription, the use of professional officers and chain of command organization; chariot, cavalry, weaponry, cross-bow and armor use and pre-figured strategic tactics employed and 2) the general literary style and authorship of Chinese texts and their materials and script used and reference material of the times; authoritative sinologists give a more plausible text dating sometime later than c. 515 BCE and between 400-300 BCE.

The Art of War is cited as originally having many more than the usual thirteen chapters and the roughly six thousand Chinese characters of the typically translated extant text.

In 1972, a partial copy of Sun Tzu's *The Art of War* written on bamboo strips along with several previously unknown texts, including the previously lost Sun Pin/Bin's *The Art of War*, are found in two Yin-Que-Shan/Silver Sparrow Mountain excavated tombs in Shandong Province dating from c. 140-118 BCE, which is considerably later than the extant text in use today.

Several hundred commentaries of Sun Tzu's *The Art of War* have been made throughout the ensuing centuries, but eleven authoritative commentaries are typically cited, most notably the first and most well-regarded of which is made by Ts'ao-Ts'ao/Cao-Cao (155-220 CE), reportedly one of the most distinguished military leaders in Chinese history.

The first Western translation of *The Art of War* is one in French made by Jean-Joseph Marie Amiot (1718-1793 CE) in 1782. The first English language partial translation is *The Book of War* made by E.F. Calthrop (1876-1915 CE) in 1905 and revised in 1908. The most popular English language translation is that of Leonard Giles (1875-1958 CE), made in 1910, who includes sharp criticisms of both previous works. His translation includes a history of Sun Wu/Tzu and *The Art of War* text, notes on the eleven commentators, references to an appreciation of Sun Tzu and an apology for war.

For purposes of the present book, there is little need for conclusive or definitive clarity on matters of authorship and text dating. The reality is that we have a tangible and reasonably coherent classic text (as with Lao Tzu's *Tao Te Ching*) that has come to us through the ages and has been, and still is, of practical value and contemporary use in creating, fashioning, experiencing and sharing the art and Way of what it is and what it means not only to be a wise and compassionate, strategic and skillful and effective and successful leader and conflict resolver and winner, but also to be a true human being and to live a real human life with fellow human beings.

THE THEMES

Contrary to what the title may seem to indicate, *The Art of War* is about dealing with and resolving conflicts, defeating enemies and winning wars strategically and tactically without fighting with an opponent or warring against an enemy through direct confrontation, engagement and combat. The objective of conflict and war is to quickly win a peace that does not deplete, drain and waste economic resources; devastate, destroy and demolish property and impoverish, annihilate and kill human beings.

Effectively and successfully dealing with and resolving conflicts involve:

1) Accepting that human conflict is an unfortunate and inevitable part of human life.

2) Having insight into human nature, the martial mentality and the military mind-set.

3) Considering conflict as a practical philosophical, psychological and social problem.

4) Knowing philosophy of conflict and war and psychology of opponents and enemies.

5) Doing pre-planning, having forethought and making calculations and deliberations.

6) Making careful, thoughtful advance assessments, evaluations, plans, preparations.

7) How to defeat enemies, achieve victories and win wars without fighting and battling.

8) Enlightened strategies, tactics, positions, dispositions, maneuvers and operations.

9) Keeping strategies and tactics unknown, unfathomable, unassailable and irresistible.

10) Knowing opponent's intent, motives, plans, configurations, positions, movements.

11) Knowing opponent's strengths and weaknesses, vulnerabilities and opportunities.

12) Knowing when, where, how to fight and not to fight re: advantages and disadvantages.

13) Preventing conflicts before they even begin and rapidly resolving them when they do.

14) Strategically using deception, dissimulation, artifice to mislead, misdirect opponents.

15) Strategically using ploys, tricks to outthink, outwit, outfox, outmaneuver opponents.

16) In direct engagements, avoiding the strengths, attacking the weakness of opponents.

17) Not fighting if and when there is no chance of winning or no advantage to be gained.

18) Catching opponents off guard by unexpected, unanticipated and unprepared for moves.

19) Weakening opponents by dislocating, destabilizing, disorganizing and dispiriting them.

20) Forcing and moving opponents into weak, disadvantageous and vulnerable positions.

21) Discouraging, disheartening, demoralizing opponents and making conflicts undesirable.

22) Changing the attitudes of opponents, inducing compliance and provoking surrender.

23) Confusing and overwhelming opponents and not overpowering, overthrowing them.

24) Moving opponents into disadvantageous positions of retreat, surrender and defection.

25) Moving opponents into positions of giving out, giving in, giving over and giving up.

26) Not using militaristic, brute force and might when engaging in conflicts with opponents.

27) Not conquering, vanquishing, demolishing, destroying and annihilating opponents.

28) Making strategic tactics flexibly adaptable to changing situations and circumstances.

29) Taking safe, favorable advantages; avoiding dangerous, unfavorable disadvantages.

30) Gaining strategic advantage and achieving tactical superiority through best positioning.

31) Using Tao's presence, power and virtuosity/te, yin-yang dynamics and wu-wei kinetics.

32) Using strategic tactics and skillful means to deflect, deter and avoid a direct conflict.

33) Being wise, calm, detached, kind, compassionate and courageous in conflict situations.

34) Considering conflict with opponents and wars with enemies as a regrettable last resort.

35) Understanding losing conflicts/wars as due to the faults, errors and failures of leaders.

36) Ending conflict with opponents as quickly as, and with as few losses as, is possible.

37) Being open to using diplomacy, negotiation, compromise, arbitration to end conflict.

38) Not only considering Tao/The Way/Art as strategies to win wars and resolve conflicts.

39) Considering Tao/The Way/Art as a peaceful Way of being a real and true human being.

THIS RENDITION

Various translations of the thirteen chapters of *The Art of War* are either continuous narratives or are arbitrarily divided into differing numbered or unnumbered paragraph sections. In this rendition, the contents of each chapter are divided into the sensible and readable unnumbered paragraphs of a narrative.

This rendition is sometimes a direct translation of the original Chinese characters, using dictionaries and bilingual translations, and is, at other times, a rendered paraphrasing of content without altering its essential meaning.

THE SYNOPSES

A 'synopsis' is defined as 'A condensed statement or outline. An abstract'. 'Synoptic' is defined as 'A general view of a whole. Manifesting or characterized by comprehensiveness or breadth of view'. Here, the synopses are not appreciably shorter and radically condensed abridgements of the thirteen chapters of the text nor are they outlined abstractions of their main concepts. Rather, they are streamlined reiterations of the content.

Almost every sentence of *The Art of War* text has a meaning that does not typically allow for its selective elimination. So, in this synopsis, content is typically reiterated in sentences that are usually made in one line. This facilitates reading them in a streamlined single horizontal form that facilitates understanding, reflecting upon and remembering their meaning. 'Synopsis' is still an appropriate designation in terms of being 'synoptic', in the sense of 'comprehensiveness or breadth of view'. Essentially, I like using the word 'synopsis' since, etymologically, it means 'viewing together'.

The Commentaries

The commentaries for the thirteen chapters are psychologically- and psychosocially-oriented and are applied to both intrapersonal/intrapychic and interpersonal/social relationships, oppositions and conflicts.[2]

Intrapersonal or intrapsychic relationships, oppositions and conflicts are those internal dualistic ones within our whole ego-Self and between our principal integral executive ego-self/I and our various differential subordinate ego-states or sub-personalities/me-s and between them and their various positive-negative traits and values, self-images and self-concepts, thoughts, feelings, actions, states and conditions, e.g.,

1) I-me, subject-object and conscious-unconscious; mind-body, head-heart, ego-Higher Self, self-Soul and body-Spirit; body and gender dysphoria; poor self-image and low self-esteem; self-blame, hatred and harm; suicidal ideation and attempts; split and multiple personalities and dissociated identities; alter-egos and ideal-selves and *doppelgangers,* etc.;

2) dualistic intrapsychic relationships, experiences, objects, contents and phenomena within consciousness that are characteristic of the aura and etheric energy body; out-of-body experiences (OBE) and near-death experiences (NDE); lucid dreaming and daydreaming; self- imaginings and fantasies; memories and reminiscences of and reflections upon past self-and life-experiences; recollections of past-lives and identities; etc.;

3) medical conditions, e.g., menstrual and pregnancy com-

plications; cancer; cardiac, cardiovascular and organ disease; diabetes and insulin resistance; physical disabilities; hereditary and congenital diseases; physical, mental and psychosomatic illness, e.g., psychoses, bipolar disorder and neuroses; psychiatric and psycholgical conditions, e.g., PTSD; personality, cognitive and behavioral disorders; mood and emotional disturbances; PMS; etc. and

4) DNA and gene abnormality; auto-immune and neurodegenerative disease; ageing; endocrine, hormonal and metabolic disorders; compromised immunity; cholesterol, fatty acid, ph imbalances; mitrochondrial deficiency, oxidative stress, glycation, free radical damage, systemic inflammation; rheumatoid arthritis, lupus, fibromyalgia, chronic fatigue, back pain, insomnia, migraines, rashes, chronic stress, etc. and

5) disease-, bacteria-, fungus- and parasite-fighting antibodies; white blood-cell leukocytes; monocytes and neutrophils; lymphocytes, b-, t- and natural killer cells and micro- and macrophages that comprise the immune system and that recognize, scavenge, engulf, destroy, absorb and remove attacking, invading and occupying pathogens in the body.[3]

Interpersonal relationships, oppositions and conflicts are those external dualistic ones between our whole ego-Self and opposing whole ego-Selves involving hierarchical role-differentiations, status differentials, negative objectifications and power struggles, e.g.,

1) within family and marital, peer and collegial and authority relationships and between I-you, I-her, I-him and I-them and self-other, e.g., parent-child, wife-husband, siblings, teacher-student, supervisor-supervisee, trainer-trainee, mentor-mentee, advisor-advisee, clergy-believer, doctor-patient, counselor-client, lawyer-client, employer-employee, etc.;

2) any and all kinds and forms of toxicity, aggression, assaulting, attacking, violence, abusing, molesting, harming, injuring, terrorizing, threatening, intimidating, harassing, groping, bullying, mugging, robbing, e.g., murdering, raping, unconsenting

sex, stalking, voyeurism, kidnapping, confining, identity theft, slandering, blackmailing, posting fraudulent social media videos, fake text messaging, etc. and

3) those involving eminent domain and evictions; capitalistic property, land and water grabs by wealthy investors, realtors and developers that, e.g., rob and displace indigenous native peoples; create dynasties, empires and monopolies; raise and fix real estate market values that price-out lower and lower middle-class human beings, create gentrification, etc..[4]

Interpersonal relationships can also include those with divine beings, avatars, spiritual masters and teachers; friendly and malevolent spirits; ghosts and apparitions; *poltergeists* and skin-walkers; manes and souls of deceased human beings and extraterrestrial and non-human beings and entities.

Non-Personal relationships, oppositions and conflicts are the I-it, I-this, I-that, I-these and I-those external dualistic ones between our whole ego-Self, Nature and the atmospheric and topographic environment; ecosystems and natural habitats; external conditions, situations and circumstances and the happenings, occurrences, events and non-human things within them, e.g.,

1) UFOs/UAPs, meteorites, solar flares, geomagnetic storms, natural disasters, global warming, climate changes, prolonged heat waves, atmospheric rivers, tornados, cyclones, typhoons, hurricanes, blizzards, heavy rainstorms, hailstorms, lightning strikes, earthquakes, avalanches, droughts, wildfires, volcanic eruptions, tsunamis, deluge, floods, landslides, erosion, melting glaciers, rising sea levels, etc. and

2) those caused by and involving non-humane and inhuman impersonal relationships, e.g., deforestation; fossil fuel use and heavy carbon footprint; CO_2, greenhouse gas and methane emission; plastic pollution and microplastics, oil spills, environmental contamination, nuclear waste; chemtrail spraying and toxic insecticide, pesticide, herbicide use; fluoridation, fake recycling; etc..

Non-Personal relationships and conflicts are also natural, unnatural and/or iatrogenic and human-induced external events, activities and substances that have detrimental and injurious intrapersonal and interpersonal effects, e.g.,

1) accidents, fractures, paralysis, burns, comas, amputation, sepsis, medical-surgical error; drug overdoses; pandemics; infectious diseases, germs, bacteria, viruses, parasites, fungus; allergens, mold; volcanic ash, toxic gases, particulate matter, second-hand smoke, asphyxiation; communicable and sexually transmitted diseases; air, blood and food-borne pathogens; radiation, hazardous waste; toxic, poisonous and lethal chemicals, PFAs/PFOs/PFOAs, 'forever chemicals', BPA, phthalate, plasticizers, heavy metals, etc.,

2) bioengineered food, GMOs; food additives, cellulose, preservatives, stabilizers, emulsifiers, dyes; seasonings, MSG, flavor enhancers; nutritional supplements and vitamins; highly processed, fried, inflammation-producing and carcinogenic foods and ingredients; red and cured meat; farm raised fish and seafood; grain-fed, hormone and antibiotic-laden livestock and poultry; hydrogenated oils, saturated and transfats; refined white flour, grains and carbohydrates; high glycemic index foods, sugar, high fructose corn syrup and artificial sweeteners; salt, etc.,

3) addictive drugs and substances; opiates, intoxicants, inhalants; prescription and over-the-counter medications; antibiotics and hormones; injections, vaccinations, anesthetics; emotional mood-altering stimulants, sedatives, analgesics, pain-killers, sleep aids, anti-depressants, inhibitors, suppressants, tranquilizers; consciousness-altering psychoactive substances and agents, etc. and

4) a poor diet of processed, non-nutritional and junk 'foods'; sedentary life-style and lack of exercise; chronic stress, poor sleep and lack of relationships all of which do not support health, promote growth, mitigate illness, sustain life; increase energy, creativity, productivity, harmony and intimacy and reduce conflict.[5]

Transpersonal relationships are:

1) non-dual, unopposed and unconflicted and intersubjective and interhuman I-Thee/we/us ones between fellow human beings and kindred human souls and between ego-self, True Tao-Nature, Higher and Deeper Tao-Self and Human Spirit.

2) ones with the Multiverse, Cosmos, Galaxy, Solar System, Planet, Heaven, Earth, Gods and Goddesses and Divine Beings, Father Sky, Mother Earth, Nature and its many phenomena and creatures and

3) ones that integrate and harmonize our personal-individual-acquired and cosmic-universal- primordial Ch'i energies and that unify and balance our physical and etheric bodies; our surrounding aura's vibrational frequency, emanation and luminous radiation and our human Soul and Spirit.

These transpersonal relationships are not objectively and structurally dualistic ones, but are subjectively and intersubjectively unified and integrated phenomenological and non-dualistic ones, and can be greatly and deeply and intensely and intimately intrapersonal and interpersonal in their nature, inter-relationships, significance, value and meaning.

Intrapersonal, interpersonal, non-personal and transpersonal relationships are all very personal ones that serve to:

1) organize, interrelate, harmonize our executive and co-equal subordinate ego-selves.

2) integrate, unify, secure, solidify, stabilize, regulate and maintain our whole ego-Self.

3) intimately interrelate our co-existing whole ego-Self and that of fellow human beings.

4) presence, safeguard, preserve, maintain, sustain and share the radiant light of our awakened, illuminated and enlightened pure consciousness, clear awareness, unique being, True Tao-Nature and Deeper and Higher Tao-Self and that of co-equal fellow human beings.

5) Soulfully and Spiritually harmoniously integrate and unify our personal Ch'i energy with the ubiquitous universal matrix and interconnected web of all-pervading cosmic Ch'i energy.

6) allow us to experience our absolute, ultimate, essential, universal, natural and innate human birthrights of enlightened consciousness, freedom, peacefulness, intimacy and happiness and

7) vitally, clearly, deeply, peacefully, freely, intimately, fully and happily create, embody, ensoul and inspirit our blessed, unique and precious true human being and real human life as one onto-bio-psycho-socio-spiritual integral whole.

THE REFERENCES

The following are references made in the text and commentaries:

In the text – the kingdom, empire, country, nation and state are referred to as the 'State'.

In the commentaries – the state is referred to as the 'Tao-State' of our pure, empty and awakened consciousness and clear, open and enlightened conscious awareness.

In the text – the emperor, king, lord, sovereign and ruler are referrred to as the 'Ruler'.

In the commentaries – the ruler is referred to as our 'True Tao-Nature'.

In the text – the military commander-in-chief and commanding general are referred to as the 'Commanding General', 'General' or 'Leader'.

In the commentaries – the leader is referred to as our principal integral 'executive ego-self'.

In the text –the military battlefield and general field are referred to as such.

In the commentaries – the field is referred to, intrapersonally, as our 'field of conscious awareness'; interpersonally, as the context, situation, nature, qualities and mood of the conflicted relationship and, non-personally, as the context, atmosphere, ground, environment, surroundings and circumstances of the conflict.

In the text – the army, military troops and fighting forces are referred to as the 'army', composed of the General, military officers and soldiers, those of both one's own and the enemy.

In the commentaries — the army is referred to as our 'whole ego-Self', composed of our principal executive ego-self and our subordinate differentiated ego-selves or subpersonalities and those of both our whole ego-Self and that of an opponent.

In the text – the military soldiers are referred to as the 'soldiers', those of both one's own and the enemy.

In the commentaries — the soldiers are referred to as our differentiated 'subordinate ego-selves' or 'sub-personalities' of our whole ego-Self and that of an opponent.

In the text – advancing, engaging, aggressing, attacking, fighting, combating, battling and warring are referred to as 'conflict' and 'conflicting'.

In the commentaries – such activities are generically referred to as 'conflict'.

In the text – environment, lands, terrain and ground are referred to as such.

In the commentaries – ground is referred to as the relationship context, medium, conditions and situations our whole ego-Self's executive and subordinate ego-selves *vis a vis* those of an opponent.

In the text – plans, strategies and tactics, formations and positions, dispositions and maneuvers, advancing and retreating and attacking and defending are referred to as such.

In the commentaries – these are all referred to as intentions, organizations and actions of our whole ego-Self's executive ego-self and subordinate ego-selves *vis a vis* those of an opponent.

In the text – weapons and weaponry are referred to as such.

In the commentaries – weapons are referred to as the force and strengths, skills and abilities, means and methods, strategies and defenses, movements and operations and options and alternatives of our whole ego-Self's executive ego-self and our subordinate ego-selves and those of an opponent.

In the text – engaging and conflicting with, defending and advancing, aggressing against and attacking, contending, fighting, combating, battling and warring with opposing, adversarial, hostile and antagonistic enemy military forces are referred to as 'conflicts' with the 'enemy'.

In the commentaries – all of the above are generally referred to: 1) as intrapersonal 'oppositions' and 'conflicts' between our executive ego-self and our subordinate ego-selves and between our subordinate ego-selves and 2) interpersonal 'oppositions' and 'conflicts' between our executive ego-self and our subordinate ego-selves of our whole ego-Self and that of an opponent.

In the text – the 'enemy' is referred to as such and as above.

In the commentaries – the enemy is contextually referred to as: 1) the intrapersonal 'others' of our whole ego-Self in opposing and conflicting relationships between our subordinate ego-selves/ me's and with our executive ego-self/I and 2) the interpersonal 'others' of opposing and conflicting relationships between our whole ego-Self and an opposing whole ego-Self. The 'enemy' is generically referred to as opposing and conflicting subordinate ego-selves, executive ego-selves and whole ego-Selves.

In the text – the Art/Tao/Way of human warfaring refers to and involves assessments and calculations, plans and preparations, strategies and tactics, positions and dispositions, actions and maneuvers and foresight, observations, knowledge and foreknowledge that result in defeating enemies and winning wars without directly fighting them and thus not violating the wholeness and depleting the resources of the State and without impoverishing, disrupting, injuring, and destroying and ending the lives of people and soldiers.

In the commentaries – the Tao/Way/Art of human Wayfaring refers to and involves the same characteristics as above with respect to the Tao-State and our True Tao-Nature and our whole ego-Self, our executive ego-self and our subordinate ego-selves that ultimately result in transpersonally overcoming and resolving intrapersonal, interpersonal and non-personal conflicts without directly 'fighting' with them and thus not violating the absolute reality, ultimate source and essential resources of our original and inborn True Tao-Nature and Tao-Virtue and the viability, vitality, vibrancy, integrity, harmony and community of our whole ego-Self and human being and that of fellow human beings.

In reading through the commentaries:

A human being *is* a 'whole ego-Self',
is True-Tao-Nature /a Higher Self
and *is* a lower executive ego-self
and lower subordinate ego-selves
and includes any and all of them.

True Tao-Nature /a Higher Self
has a lower executive-ego
and lower subordinate ego-selves
and essentially *is not* any of them
but may be identified with them.

The lower executive ego-self
has lower subordinate ego-selves
but existentially *is not* them
but may be identified with,
or disidentified from, them.

YOUR REFLECTIONS

TAO TE CHING
PASSAGES

The spirit, philosophy, psychology and conduct of conflict and war and its involved strategic tactics as presented in Sun Tzu's *The Art of War* are rooted in the spiritual tradition of ancient Chinese Taoism.

Lao Tzu, the 'founder' and early articulator of Chinese Taoism, disheartened by the political intrigue and hegemonic feudal conflicts of the late Spring and Autumn Period and nearing the internecine warfare and carnage of the early Warring States Period of Chinese history (c. 6th-5th Centuries BCE), ultimately leaves China and heads for the K'un-Lun mountains of immortality.

On Lao Tzu's passage out of the country, he is asked by the keeper of the pass, Yin-Hsi, to record his wisdom that later becomes known as the *Tao Te Ching*.

The following are passages from Lao Tzu's *Tao Te Ching*, a principal text of Taoism, regarding matters of power, force, aggression, possessiveness, contention, conflict, fighting, weapons and war.

'Most developed goodness is being like water. Water is benefiting all beings without contending...We are not contending and there is no harming'. *TTT 8.*

'According with Tao is being everlasting and free from endangering throughout our lifetime'. *TTC 16.*

'As wise human beings, we are embodying One and being world models. Not displaying...not asserting...not boasting... not parading...not contending with anyone'. *TTT 22.*

'Desiring to take over our world, trying to control and manipulate it...we cannot be succeeding. Our world is a sacred vessel not to be interfered with. Acting upon it is ruining it. Holding onto it is losing it'. *TTT 29.*

'Using Tao in assisting leaders is not governing our world by forcing. Using force is recoiling on its users. Where forces are dwelling, brambles are growing. Where forces are fighting, famines are following. When some force must be used to achieve necessary results, it is used only reluctantly and stopped without priding, boasting, parading or dominating human beings. Overdeveloping power is accelerating decay. What is not following Tao is quickly coming to an early ending'. *TTC 30.*

'Weapons are instruments of disaster. As most developed human beings, we are deploring them and their use. Embodying Tao, we are living without them...Should we be required to use them, it is as a last resort and with utmost restraint. Even if victorious, as most developed leaders, we are not enjoying it. Enjoying victory is enjoying slaughtering. Enjoying slaughtering, we are not succeeding in the world. We are honoring...the right at unfortunate events...Chief generals are standing on the right, an arrangement like that of a funeral. Slaughtering multitudes of human beings is bringing grief and sorrow. We are treating victories as funerals'. *TTC 31.*

'Understanding others, we are being intelligent. Understanding ourselves, we are being illuminated. Overcoming others, we are being outwardly powerful. Overcoming ourselves, we are being inwardly strong...Using forceful power, we may be achieving objectives but connecting with Tao, we are enduring'. *TTC 33.*

'Fish are not leaving deep waters. We are not displaying powerful instruments of the State'. *TTC 36.*

'As ancient ones are teaching, so am I teaching once again. As aggressive violent people, we are coming to violent endings. This is my essential teaching'. *TTC 42.*

'When Tao is present in our world, race horses are working in fields. When Tao is not present in our world, war horses are breeding at shrines. No misfortune is greater than wanting what others have. No disaster is greater than wanting to have more. No calamity is greater than not having enough. Being content with what one has, knowing when enough is enough, is constantly having enough'. *TTC 46.*

'Our world is being gained by non-interfering. When we are interfering, our world is being lost'. *TTC 48.*

'As wise human beings, we are living in one human world and harmoniously uniting the heart-minds of human beings..... regarding them as our own children'. *TTC 49.*

'Cultivating life well...we are not encountering soldiers or weapons in wars, having no place for...swords of soldiers...there is no place for dying'. *TTC 50.*

'The Great Pathway is very straight, yet human beings are taking detours...ornate weapons are being displayed, storehouses are being depleted...surplus riches are being hoarded. All this is vanity and thievery and certainly not cultivating Tao'. *TTC 53.*

'We are governing this State straightforwardly. We are deploying armies strategically. We are gaining our world by not interfering...through Tao...more weapons, more disordered countries'. *TTC 57.*

'We are acting without forcing...addressing difficult issues early while they are still being easy, addressing great matters while they are still being small...As wise human beings, we are not striving for greatness, yet are often attaining greatness'. *TTC 63.*

'We are anticipating issues before they are coming into being. We are regulating matters before they are becoming out of order...Forcing things is ruining them. Seizing things is losing them'. *TTC 64.*

'We are having three treasures...Unconditioning love, we can be courageous. Conserving resources, we can be generous. Restraining precedence, we can be splendorous...Being without them, can bring ruin and death. Through unconditioned loving, we are succeeding in the offensive, sustaining in the defensive. Heaven is protecting and saving us through unconditioned loving'. *TTC 67.*

'As most developed warriors, we are not being hurtful. As most developed fighters, we are not being rageful. As most developed victors, we are not being vengeful. As most developed leaders, we are not being forceful. This is the Virtuosity of not contending, the potency of not coercing and matching the ultimacy of heaven'. *TTC 68.*

As military strategists, we are not taking the offensive and are taking the defensive. We are not advancing an inch and are retreating a foot. This is deploying forces without marching them, facing opponents without engaging them, displaying weapons without employing them, defeating armies without battling them. There is no greater calamity than creating and fighting 'enemies' and losing our three treasures. So, when opposing forces are warring, only unconditioned loving is winning'. *TTC 69.*

'Being courageous by daring is risking dying. Being courageous by not daring is preserving living.' *TTC 73.*

'Hard and rigid are accompanying dying. Soft and fluid are accompanying living. So, fixed armies will be shattered'. *TTC 76.*

'As great grievances are being reconciled, some discontent is usually remaining. How is this being made good? As wise human beings, we are owning our part in matters without blaming the other parties. Being with Virtuosity, we are fulfilling agreements. Being without Virtuosity, we are demanding reciprocations'. *TTC 79.*

'Here is a small State...As human beings...though there are some weapons, we are not displaying them...living simply and sufficiently...being delighted with our usual ways...we are happily growing old and dying'. *TTC 80.*

'Developed human beings are not always arguing. Arguing people are not always developed...Heaven's Tao is benefiting, not harming. Wise human beings' Tao is assisting, not contending'. *TTC 81.*

THE ART OF WAR
CHAPTER QUOTES

1. 'War is of vital importance to the State; a serious matter of life or death and the road to survival or destruction. It is a subject of investigation and examination that cannot be ignored or neglected.....All war is fundamentally and essentially the art of deception and deceiving the enemy'.

2. 'Protracted campaigns will deplete the State's resources..... skillfulness has not characterized prolonged wars and no country has ever benefited from them'.

3. 'To fight and conquer in all battles is not supreme excellence. Supreme excellence consists in strategically breaking the enemy's resistance and subduing them without engaging and fighting.....Knowing the enemy and oneself, victory results in a hundred battles'.

4. 'Good fighters skillfully place themselves beyond the possibility of being defeated and wait for vulnerable opportunities to defeat the enemy. Securing against defeat is one's own doing and defeating the enemy relates to their doing.....the certainty of victory is conquering an enemy that is already defeated.....the victor only seeks a battle after the victory has already been won'.

5. 'Simulated disorder, fear and weakness require order, courage and strength based upon organization, energy and strategic tactical maneuvering'.

6. 'Irresistable advances can be made, if the enemy's weak points are gone for. Safe evasion of their pursuits can be made if elusive movements are quicker than theirs'.

7. 'The enemy Commanding General and a whole enemy army can be made to lose presence of mind, heart, spirit and energy'.

8. 'The *Art of War* is not relying upon the likelihood of the enemy not arriving but rather upon being ready to receive them and not upon the chance of them not attacking but rather upon making positions unassailable'.

9. 'The General who has no forethought and underestimates the enemy will be captured by them'.

10. 'When military soldiers are regarded as one's own beloved sons, they will follow into the deepest valleys and stand by to the death'.....When you know the enemy, yourself, heaven and earth; victory will be certain and complete'.

11. 'Initially, be shy like a young maiden until the enemy gives you an opening and then spring forward with the speed of a darting rabbit and it will be too late for the enemy to resist'.

12. '.....a State once destroyed can never be restored nor can the dead ever be revived. So, the enlightened Ruler and wise Commanding General are alert and attentive to this and are prudent and cautious. This is the way to keep a State secure and at peace and to preserve an army whole and intact'.

13. 'Raising large armies and marching them great distances entails heavy losses for the people and depletes the resources of the State.....only enlightened Rulers and wise Commanding Generals who use the greatest intelligence of spying and espionage will achieve the greatest success. Relying upon spies and secret spying operations is a most essential factor in the *Art of War*, because upon them depends an army's every action.

Your Reflections

THE PLAY OF CONFLICT AND WAR

THE PLAY. The Way to Win Intrapersonal and Interpersonal Conflicts and Wars.

THE SETTINGS

1) In China during the Eastern Chou/Zhou Dynasty (770-256 BCE) and the late Spring and Autumn Period (770-475 BCE) and the early Warring States Period (475-221 BCE) when feudal states were vying for hegemony.

2) Any time, place and way when, where and how opposing whole ego-Selves are engaged in intrapersonal and/or interpersonal power struggles, conflicts and wars.

THE CAST

1) The Lord, Ruler, Emperor of a feudal state aka our True Tao-Nature, Tao-Being, Pure Consciousness, Transcendental Ego, Higher/Deeper/Transpersonal Self and Body-Spirit/Soul of the State of Tao.

2) The Military Commander-in-Chief or Commanding General aka our leading and principal, executive and consciously aware ego-self of our whole ego-Self and that of an opponent.

3) The Army, troops, soldiers aka our subordinate ego-selves and sub-personalities of our whole ego-Self and those of an opponent.

4) The Opposition — an opposing feudal State, ruler, general, army and soldiers aka an opposing whole ego-Self and its executive ego-self and subordinate ego-selves engaged in conflict.

THE STAGE

The battlefield aka our field of consciousness, our personal world of intrapsychic relationships, our social world of interpersonal relationships and our environmental world of non-personal relationships.

THE PROPS

Weapons, arms aka the ego-self strengths, abilities, skills, defenses, strategies, tactics, etc. of our executive ego-self and our subordinate ego-selves of our whole ego-Self and those of an opponent.

THE SCRIPT

1) Defeat the opposition, resolve the conflict and win the war without draining resources or losing personnel through foreknowledge and understanding human nature, True Tao-Nature and the weakness and strengths of whole ego-Selves;

2) Plan strategies, tactics and operations ahead of time; assess advantages and disadvantages of fighting and anticipate and manage contingencies;

3) Know the atmosphere and terrains of situations; devise and contrive strategic tactical maneuvers and variations and adapt flexibly to changing conditions;

4) Attack weaknesses of the opposition and avoid their strengths;

5) Use deception and surprise to outfox and demoralize the opposition;

6) Make rapid and unexpected maneuvers, advances and attacks that the opposition cannot prepare for and

7) Employ intelligence-gathering go-betweens or spies and secret espionage agents.

All of the above are designed and conducive to effectively deal with and to successfully resolve conflicts by defeating the opposition and ending the war.

THE ROLES

1) *The Ruler*, aka our True Tao-Nature, is accorded with the Tao/the Way, humanely shares the one will of the people and originates the decision to deal with and to resolve conflicts; lends positive energy to, presides over and witnesses all ego-operations and holds them, the country, the Commanding General and the military officers and soldiers in heart.

2) *The Commanding General*, aka the leader and our principal executive and consciously aware ego-self of our whole ego-Self, is in charge of collecting, organizing, encamping and regulating our subordinate ego-selves and ordering, commanding, conducting and executing all strategic and tactical operations, i.e., structuring ranks and chain of command, delegating authority and responsibilities, giving out orders and equitably managing, controlling, motivating, rewarding, disciplining, positioning and deploying our subordinate ego-selves of our whole ego-Self.

3) *The Soldiers,* aka our subordinate ego-selves of our whole ego-Self, loyally follow commands and orders, dutifully implement strategies and tactics and willingly engage and fight an opposing whole ego-Self and maintain a readiness to die for the cause without retreating from the opposition and deserting the Tao-State, our Tao-Nature, our whole-ego-Self, our executive ego-self and its other subordinate ego-selves

4) *The Opposition* – A) in Act One; dissatisfied, discontented, disgruntled, unruly, dissenting, demoralized, insubordinate and mutinous opposing soldiers aka our subordinate ego-selves vying for favor, rewards, promotion and superiority; B) in Act Two; an opposing ruler, commander and army of hostile soldiers from another feudal country vying for hegemony aka an executive ego-self and subordinate ego-selves of an opposing whole ego-Self intending to encounter, advance upon, confront, attack, engage, fight, battle, defeat, win, conquer, vanquish, etc.; C) in Act Three; the atmosphere and topography of the opposing ground, terrain and situation are the natural surroundings

and environmental conditions that are either safe or dangerous and advantageous or disadvantageous to our whole ego-Self and D) in Act Four, conflicts with opposing whole ego-Selves are transpersonally anticipated, effectively planned for, successfully dealt with and completely resolved by, through and across the use of intermediary go-betweens and secret intelligence-gathering spies and double espionage agents.

ACT ONE. INTRAPERSONAL CONFLICT – the I-me duality within our whole ego-Self.

Intrapersonal conflicts within our whole ego-Self between our dualistically objectified and opposed: 1) subjective 'I' and objective 'me' ego-self, 2) our principal executive ego-self and our numerous subordinate ego-selves and 3) between our numerous subordinate ego-selves and including opposing and conflicting individual traits and dispositions, sensations and perceptions, thoughts and feelings, actions and activities, identities and identifications, etc..

Scene One ❖ *Making Plans.*
Scene Two ❖ *Waging War.*
Scene Three ❖ *Strategic Attack.*
Scene Four ❖ *Military Forms.*

ACT TWO. INTERPERSONAL CONFLICT — the self-other duality between our whole ego-Self and that of the opposition.

Interpersonal conflicts between a dualistically objectified and opposed: 1) our whole ego-Self and an opposing whole ego-Self, 'me' and 'you'/'her'/'him'/'them'; 2) our executive ego-self/'I' and an opposing executive ego-self and 3) our subordinate ego-selves and opposing ego-selves and including opposing and conflicting characteristics, motives and intentions, plans and purposes, strategies and tactics, goals and objectives, advantages and disadvantages, activities and movements, strengths and weaknesses, skills and abilities, etc..

Scene Five ❖ *Soldier Power.*
Scene Six ❖ *Weakness and Solidity.*
Scene Seven ❖ *Military Fighting.*
Scene Eight ❖ *Nine Variations.*

ACT THREE. NON-PERSONAL CONFLICT— the I-it duality between our whole ego-Self and the environment.

Non-personal or impersonal conflicts are related to: 1) the dangers and disadvantages of; the safe defensive and risky offensive positions and maneuvers within and the atmospheric conditions and topographic configurations of the land, ground, terrain and environment of the conditions, situations and circumstances of conflicts and 2) the impersonal incendiary attacks upon the executive ego-self and subordinate ego-selves of an opposing whole ego-Self as non-entities and anonymous ciphers and upon their supplies, resources, communications and lines of reinforcements.

Scene Nine ❖ *Army Conduct.*
Scene Ten ❖ *Terrain Forms.*
Scene Eleven ❖ *Nine Situations.*
Scene Twelve ❖ *Fire Attacks.*

ACT FOUR. TRANSPERSONAL CONDUCT — the I-Thou bipolar unity, by through, across and beyond our whole ego-Self and that of the opposition.

Transpersonal conflicts: 1) within our whole ego-Self, 2) between our executive ego-self and our subordinate ego-selves of our whole ego-Self and 3) between an opposing and conflicting whole ego-Self are being handled, resolved, integrated, unified and transcended through co-ordination and co-operation and through obtaining and utilizing information, intelligence, foreknowledge and advantages gained through, by, from and across intermediary go-betweens, i.e., inconspicuous spies and secret espionage and double agents who are involved in the unobtrusive, surreptitious, clandestine and stealthy activities of infiltration, espionage, subterfuge and subversion.

The co-existing interchanging and exchanging by, through and across our subordinate ego-selves, our executive ego-self, our whole ego-Selves and our True Tao-Nature/Pure Consciousness/ Higher and Deeper Self are essential: 1) to the actions, interactions and transactions; 2) to the viability, vitality and integrity and 3) to the safety, survival and success our whole ego-Self in conflicts with an opposing whole ego-Self.

Thus, the original, absolute and essential unity and the primordial, acquired and available energy of the ultimate State of Tao and our inborn True Tao-Nature are safeguarded and preserved; reserved and conserved; harmonized and regulated and peaceful and fulfilled.

Scene Thirteen ❖ *Using Go-Betweens.*

FINE

What is war really for?
Conflicts of dualistic either-or.
Learned as part of human lore.
Having less and desiring more.

What is war really for?
All of the loss, blood and gore.
Trying to settle a deluded score.
Not a part of Heart-Soul's core.
Sends humans to the far shore.

YOUR REFLECTIONS

Chapter Title
Comparisons

Ames ❖ 1. On Assessments. 2. On Waging Battle. 3. Planning the Attack. 4. Strategic Dispositions. 5. Strategic Advantage. 6. Weak Points and Strong Points. 7. Armed Contest. 8. Adapting to the Nine Contingencies. 9. Deploying the Army. 10. The Terrain. 11. The Nine Kinds of Terrain. 12. The Incendiary Attack. 13. Using Spies.

Cleary ❖ 1. Strategic Assessments. 2. Doing Battle. 3. Planning a Siege. 4. Formation. 5. Force. 6. Emptiness and Fullness. 7. Armed Struggle. 8. Adaptations. 9. Maneuvering Armies. 10. Terrain. 11. Nine Grounds. 12. Fire Attack. 13. On the Use of Spies.

Giles ❖ 1. Laying Plans. 2. Waging War. 3. Attack by Stratagem. 4. Tactical Dispositions. 5. Energy. 6. Weak Points and Strong. 7. Maneuvering. 8. Variation in Tactics. 9. The Army on the March. 10. Terrain. 11. The Nine Situations. 12. The Attack by Fire. 13. The Use of Spies.

Gagliardi ❖ 1. Analysis 2. Going to War. 3. Planning an Attack. 4. Positioning. 5. Momentum. 6. Weakness and Strength. 7. Armed Conflict. 8. Adaptability. 9. Armed March. 10. Field Position. 11. Types of Terrain. 12. Attacking with Fire. 13. Using Spies.

Griffith ❖ 1. Estimates. 2. Waging War. 3. Offensive Strategy. 4. Dispositions. 5. Energy. 6. Weaknesses and Strengths.

7. Maneuvers. 8. The Nine Variables. 9. Marches. 10. Terrain. 11. The Nine Varieties of Ground. 12. Attack by Fire. 13. Employment of Secret Agents.

Huynh ❖ 1. Calculations. 2. Doing Battle. 3. Planning Attacks. 4. Formation. 5. Force. 6. Weakness and Strength. 7. Armed Struggle. 8. Nine Changes. 9. Army Maneuvers. 10. Ground Formation. 11. Nine Grounds. 12. Fire Attacks. 13. Using Spies.

Ivanhoe ❖ 1. Assessing. 2. Waging War. 3. Offensive Strategy. 4. Disposition of Forces. 5. Strategic Potential. 6. Tenuousness and Solidity. 7. The Clash of Arms. 8. Nine Variations. 9. Maneuvering Forces. 10. Dispositions of Terrain. 11. Nine Types of Terrain. 12. Attacking with Fire. 13. On the Use of Spies.

Kaufman ❖ 1. Considerations and Estimations for War. 2. Preparations for War. 3. The Nature of Attacks. 4. How to Think of War. 5. Using the Power of Heaven. 6. Fortitude and Frailty. 7. Manipulation of Circumstance. 8. Variations of Reality in War. 9. The Virtue of Changing Positions. 10. Control and Maintenance of Territory. 11. Conducting and Managing Campaigns. 12. Fierceness in Combat. 13. Spies and Traitors.

Mair ❖ 1. Initial Assessments. 2. Doing Battle. 3. Planning for the Attack. 4. Positioning. 5. Configuration. 6. Emptiness and Solidity. 7. The Struggle of Armies. 8. Nine Varieties. 9. Marching the Army. 10. Terrain Types. 11. Nine Types of Terrain. 12. Incendiary Attack. 13. Using Spies.

Minford ❖ 1. Making of Plans. 2. Waging of War. 3. Strategic Offensive. 4. Forms and Dispositions. 5. Potential Energy. 6. Empty and Full. 7. The Fray. 8. The Nine Changes. 9. On the March. 10. Forms of Terrain. 11. The Nine Kinds of Ground. 12. Attack by Fire. 13. Espionage.

Nylan ❖ 1. First Calculations. 2. Initiating Battle. 3. Planning an Attack. 4. Forms to Perceive. 5. The Disposition of Power. 6. Weak and Strong. 7. Contending Armies. 8. Nine Contingencies. 9. Fielding the Army. 10. Conformations of the Land. 11. Nine Kinds of Ground. 12. Attacks with Fire. 13. Using Spies.

Pepper & Wang ❖ 1. Planning. 2. Waging War. 3. Plan of Attack. 4. Tactics. 5. Military Power. 6. Weakness and Strength. 7. Battle. 8. The Many Transformations. 9. Deployment of Troops. 10. Terrain. 11. Types of Terrain. 12. Attack by Fire. 13. Use of Spies.

Sawyer ❖ 1. Initial Estimations. 2. Waging War. 3. Planning Offensives. 4. Military Disposition. 5. Strategic Military Power. 6. Vacuity and Substance. 7. Military Combat. 8. Nine Changes. 9. Maneuvering the Army. 10. Configurations of Terrain. 11. Nine Terrains. 12. Incendiary Attacks. 13. Employing Spies.

Vespe ❖ 1. Making Plans. 2. Waging War. 3 Strategic Attack. 4. Military Forms. 5. Soldier Power. 6. Weakness and Solidity. 7. Military Fighting. 8. Nine Variations. 9. Army Conduct. 10. Terrain Forms. 11. Nine Situations. 12. Fire Attacks. 13. Using Go-Betweens.

Wing ❖ 1. The Calculations. 2. The Challenge. 3. The Plan of Attack. 4. Positioning. 5. Directing. 6. Illusion and Reality. 7. Engaging the Force. 8. The Nine Variations. 9. Moving the Force. 10. Situational Positioning. 11. The Nine Situations. 12. The Fiery Attack. 13. The Use of Intelligence.

Chapter Focuses

The overall and principal focus of *The Art of War* is successfully defeating an enemy and winning a war without fighting, losing soldiers and depleting resources through relevant and effective leadership, knowledge, planning, strategies, tactics, troop morale, military maneuvers and foreknowledge.

The following are the general focuses of each of the thirteen *The Art of War* chapters.

CHAPTER ONE ❖ MAKING PLANS.
Plans and Preparations. An army's calculations.

The importance of war. Five factors determining conditions and governing war. The basis for comparing armies. Forecasting victory or defeat. Characteristics of the leader. Advice regarding circumstances, rules and modifying plans. Deception. Relationships with specific characteristics of the enemy. Secrecy about military devices. Temple calculations and rehearsing plans and success or failure in battle. Foreseeing victory or defeat.

CHAPTER TWO ❖ WAGING WAR.
Expenses and Objectives. An army's directives.

The cost and funding of war, protracted campaigns and their effects. The need for cleverness. The impoverishment, substantial disruption and draining of the lives of people. Expenses for equipment, etc.. Foraging the enemy. Treatment of captured soldiers. The role of the leader.

CHAPTER THREE ❖ STRATEGIC ATTACK.
Strategies and Essentials. An army's potentials.

Capturing an army and taking a country whole. Defeating the enemy without attacking, fighting, besieging or lengthy field operations. Using strategies. Rules in war. The losing ways of a leader. A trusting army. Five essentials for victory. Knowing the enemy and oneself.

CHAPTER FOUR ❖ MILITARY FORMS.
Opportunities and Tactics. An army's logistics.

The nature of and opportunity for defeating the enemy. Defensive and offensive tactics. Excellence and skillfulness in fighting. Victorious strategy. Qualities of the leader. Military method.

CHAPTER FIVE ❖ SOLDIER POWER.
Energy and Momentum. An army's continuum.

Controlling military forces. Direct and indirect maneuvering and methods and their characteristics. Effective simulations and tactics in battle. The value and momentum of the combined energy of soldiers.

CHAPTER SIX ❖ WEAKNESS AND SOLIDITY.
Formlessness and Flexibility. An army's utility.

Characteristics of effective and ineffective strategies against the enemy. Being subtle and secret. Strong and weak points. Emptiness and fullness. Formlessness, invisibility and unitedness. Using the enemy's own tactics for defeating them. Flexible regulating, varying and modifying of strategic tactics in relation to changing circumstances and conditions.

CHAPTER SEVEN ❖ MILITARY FIGHTING.
Requirements and Maneuvers. An army's movers.

Collecting and concentrating the army. The nature, methods, conduct, requirements characteristics, timing and mood of strategic tactical maneuvering. The *Art of Warfare.*

CHAPTER EIGHT ❖ NINE VARIATIONS.
Advantages and Contingencies. An army's agencies.

Tactical conduct and advantages and disadvantages in varying situations. The teaching of the *Art of War.* Five dangerous faults of leaders.

CHAPTER NINE ❖ ARMY CONDUCT.
Positions and Dispositions. An army's conditions.
Encamping and strategically and effectively moving, positioning and conducting the army. Observing and detecting behaviors of the enemy. Signs and meanings of enemy conduct and movements. Treatment of and relationships with soldiers.

CHAPTER TEN ❖ TERRAIN FORMS.
Grounds and Vulnerability. An army's mobility.
Six kinds of terrain and conduct in relation to them. Six calamities to which an army is exposed that risk being defeated. Putting knowledge into practice. The rules of fighting. The winning and losing characteristics of leaders. Knowing Heaven-Earth, oneself, one's own soldiers and the enemy and winning wars.

CHAPTER ELEVEN ❖ NINE SITUATIONS.
Situations and Variations. An army's stations.
Nine varieties of ground. Conduct on the nine varieties of ground. Winning strategies and tactics in relation to the enemy. Knowing the predispositions and behaviors of soldiers. Qualities, behaviors and knowledge of the skillful leader and tactician. Essential, necessary and successful behavior in relation to conditions, soldiers, the army and the enemy.

CHAPTER TWELVE ❖ FIRE ATTACKS.
Fire and Leadership. An army's gamesmanship.
The five ways of attacking by fire. Materials for and timing of fires. Meeting five possible developments. The values, qualities, attitudes, behaviors and wisdom of the leader.

CHAPTER THIRTEEN ❖ USING GO-BETWEENS.
Spying and Foreknowledge. An army's edge.
Reiteration of the reasons for avoiding war. The value of the leader's foreknowledge. Five classes of spies and their use.

Maintaining intimate relationships with spies. Conduct in relation to spies. The qualities of spies. The spies of the enemy. The value and use of intelligence. The aim and goal of spying. The treatment of converted spies. The critical importance of spies for effective military movements and successful results in warfare.

CHAPTER FOCUSES

1. Plans and preparations. An army's calculations.
2. Expenses and objectives. An army's directives.
3. Strategies and essentials. An army's potentials.
4. Opportunities and tactics. An army's logistics.
5. Energy and momentum. An army's continuum.
6. Formlessness and flexibility. An army's utility.
7. Requirements and maneuvers. An army's movers.
8. Advantages and contingencies. An army's agencies.
9. Positions and dispositions. An army's conditions.
10. Grounds and vulnerability. An army's mobility.
11. Situations and variations. An army's stations.
12. Fire and leadership. An army's gamesmanship.
13. Spying and foreknowledge. An army's edge.

Intrapersonal Relationships

SUN-TZU

CHAPTER ONE
MAKING PLANS

始 計

SHIH/SHI
Begin/*Make*

NU/NU – female/woman
+
I/YI – I/oneself/cure

CHI/JI
Plan/Strategem

YEN/YAN – word/speak
+
SHIH/SHI – ten

THE TEXT

SUN TZU SAYS:

War is of vital and critical importance for the State. It is the ground and most serious matter of life or death and the road to survival or destruction. Therefore, its investigation and examination cannot be ignored or neglected.

The *Art of War* is regulated by five factors that need to be considered, calculated and assessed when determining and understanding the conditions of war and the battlefield.

1. *Moral Law* – is Tao/Way/method, the complete accord of the Ruler, the General and the people who all share the same objectives and who will follow without betrayal and irrespective of dangers and the risk of possible death.

2. *Heaven* – is Nature, yin-yang, night and day, cold and hot weather and the cycling of seasons.

3. *Earth* – is Nature, land, the terrain of small and great distances, high and low ground, wide open ground and narrow passes, safe or dangerous land formation holding the possibilities of life or death.

4. *The Commanding General* – is an awakened, enlightened and liberated Tao-Master who is an effective and successful leader embodying the virtues of wisdom, trustworthiness, humaneness, courage and discipline.

5. *Structure and Discipline* – are the organization of sub-divisions in the army, gradations of rank among officers, chain of command, delegation of authority, clear regulations and standards of discipline, reward and punishment.

Understanding these five factors will bring victory. Not understanding them will bring defeat.

When planning and assessing military operations, these five factors can also be a way of comparing countries, Rulers, Generals and armies:

1. Which Ruler embodies and enacts Tao/the Way/method to a higher degree?

2. Which General is a more identified Tao-Master and has the greater abilities?

3. Which country has more advantageous weather and terrain of heaven-earth?

4. Which army has more rigorously structured, enforced and effective discipline?

5. Which army is larger, more unified, stronger, more courageous and powerful?

6. Which army of officers and soldiers is more highly trained and experienced?

7. Which army has more clarity and consistency in rewards and punishments?

By means of these seven considerations, victory or defeat in a war can be forecasted.

Generals abiding by and acting upon these considerations

will defeat an enemy and should be retained and those who do not will be defeated by an enemy and should be dismissed.

Beyond these considerations, the General should also be aware of and utilize any and all favorable and advantageous circumstances and should organize the army and modify plans, strategies, tactics and maneuvers accordingly.

All warfare is fundamentally and essentially the Tao/Way/ Art of deception and deceiving and outwitting an enemy through artifice, subterfuge, stratagems, dissimulation, pretense, ruse, tricks, ploys, baits, etc..

For example: when able, appear unable; when ready, appear unready; when near, appear distant; when distant, appear near.

If an enemy seeks advantage, entice them, feign disorder and then subdue them.

If an enemy is solid, prepare for them. If an enemy is strong, evade them.

If an enemy is angry, irritate them. If an enemy is disorganized, seize them.

Pretend to be weak and provoke arrogance. If an army is resting, tire them.

If an enemy force is unified, divide and separate them.

Attack an enemy when and where they are unprepared and appear where and when unexpected. This is how to defeat an enemy and to win wars. Any military knowledge and strategies are not divulged ahead of time.

Generals making and rehearsing many calculations in a pre-war temple council devoted to campaign plans, strategies, tactics and maneuvers before any battles are fought are victorious. Those making few calculations are defeated and how much more so will be those making no calculations at all!

Through attending to these matters, who will win or lose a war can be foreseen.

THE SYNOPSIS

War is a grave matter of life or death, survival or destruction and cannot be ignored.

The *Art of War* is regulated by, armies can be compared by and victory can be gauged by:

1. *Tao/the Way* - complete accord. Which Ruler, General and people have the greatest?

2. *Heaven* - the weather. Which country has the most favorable weather conditions?

3. *Earth* - the terrain. Which country has the most viable and advantageous terrain?

4. *The General*. Which has the most wisdom, trustworthiness, humaneness, courage, ability?

5. *The Army*. Which has the best training, chain of command, rules, discipline and strength?

Generals abiding by and enacting these considerations are retained, if not, dismissed.

Generals use any and all advantages and modify plans, strategies and tactics as necessary.

War involves deception and feigning a lack of readiness, strength, ability, order, distance.

War involves baiting, provoking, tiring, separating, dividing and evading an enemy army.

Attack an enemy where and when unprepared and appear when and where unexpected.

Pre-planning, rehearsing and calculating are necessary for winning and not being defeated.

No knowledge, information, plans, strategies or tactics are to be disclosed ahead of time.

Through attending to these matters, who will win or lose a war can be foreseen.

THE COMMENTARY

Our True Tao-Nature is our pure and empty consciousness, our clear and open awareness and our Tao-Self and Higher/ Deeper Self.

Our whole ego-Self is the intrapersonal or intrapsychic bipolar unity of our subjective executive ego-self/'I' and our numerous objectified subordinate ego-selves/'me-s' or subpersonalities.

Our executive ego-self is our awakened, enlightened, liberated and Tao-identified 'leader' in charge of gathering, organizing, integrating, regulating, ordering, directing, mobilizing and utilizing our subordinate ego-selves, e.g., we mostly identify our 'self'/'I' with our executive ego-self rather than with our whole ego-Self which includes our numerous and often dualistic, antithetical, opposing, antagonistic and conflicting subordinate ego-selves/'me-s'.

Our various subordinate ego-selves/'me-s' of our whole ego-Self are: 1) individual objectified characteristics, qualities and traits, e.g., kindness-unkindness, humility-arrogance, neatness-messiness; 2) identifications, images and concepts, e.g., authentic-inauthentic, beautiful-ugly, sensitive-insensitive and 3) thoughts, feelings and actions, e.g., positive-negative, happy-sad, appropriate-inappropriate; etc..

Any and/or all of our subordinate ego-selves can be either integrally and harmoniously co-existing or differentially and separately opposing and conflicting, e.g., an inauthentic subordinate ego-self may seek: 1) to be superior to an authentic subordinate ego-self, 2) to usurp the authority of the executive ego-self, 3) to assume being the whole ego-Self and/or to displace the reality of our True Tao-Nature or Higher/Deeper Self.

Under various requisite circumstances, our executive ego-self can be completely identified with and as one of our subordinate ego-selves acting in maintaining the integrity of our whole ego-Self in relation to an opposing subordinate ego-self, e.g., one of our strong subordinate ego-selves may be appropriately utilized to effectively deal with, and to successfully resolve, an intrapersonal conflict with an opposing subordinate ego-self.

Conflict is of vital importance: 1) for the source, presence, integrity, consciousness, resources and manifestation of our True Tao-Nature and 2) for the viability, meaning, harmony and functionality of our whole ego-Self, i.e., the rulership of our True Tao-Nature, the leadership of our principal executive ego-self and the compliance of our numerous subordinate ego-selves.

Intrapersonal or intrapsychic opposing and conflicting relationships can obscure and displace our True Tao-Nature and can distract, disorganize, disrupt, deviate and obstruct our whole ego-Self. Therefore, such conflicts need to be seriously and carefully attended to, considered, investigated, studied, known and deliberated by us. Plans and objectives need to be devised and strategies and tactics need to be contrived to effectively deal with and to successfully resolve our intrapersonal conflicts.

The Tao/Way of intrapersonal conflict is within our whole ego-Self and is: 1) opposition between our executive ego-self and our subordinate ego-selves and 2) oppositions between our various subordinate ego-selves. Our executive ego-self is ordinarily identified with the 'I' of our whole ego-Self; e.g. our being, self-sense, presence, identity, etc..

Our various subordinate ego-selves are ordinarily identified with the 'me-s' of our whole ego-Self; e.g. true-false me, good-bad me, right-wrong me, healthy-ill me, successful-unsuccessful me, happy-sad me, calm-anxious me, intimate-lonely me, etc..

The nature, kinds, forms, degree, intensity and extent of intrapersonal conflict, either between our executive ego-self and subordinate ego-selves or between our various subordinate ego-selves, occur or do not occur according to:

1) Whether or not there is integral unity and harmonious accord of our True Tao-nature, our executive ego-self and our subordinate ego-selves of our whole ego-Self and conscious

awareness, independent of life-dangers and death-risks, e.g., in and as our enlightened or unenlightened consciousness. Some of our subordinate ego-selves may be more or less identified with our True Tao-Nature, e.g., more or less accorded with Tao and the Tao-State.

2) Whether or not heavenly yin-yang bipolar unities of, e.g., cold-hot weather, cloudy-clear skies, night-day, winter-summer, advantageous-disadvantageous climates, etc. are internalized by our whole ego-Self as context, atmosphere and conditions e.g., arguing with opposing parts of ourself on a blazing hot summer day. Some of our subordinate ego-selves may be more or less identified with our Heavenly Tao-Nature, e.g., more or less airy and spacious.

3) Whether or not earthly ying-yang bipolar unities of, e.g., areas that are close-distant, high-low, open-narrow, level-precipitous, safe-dangerous environments, etc. are internalized by our whole ego-Self as content, surroundings and situations, e.g., comparing opposing aspects of ourself in our mind, through our physical touch or as reflected in a mirror. Some of our subordinate ego-selves may be more or less identified with our Earthly Tao-Nature, e.g. more or less solid and grounded.

4) Whether or not our executive ego-self of our whole ego-Self is embodying and enacting wisdom, trustworthiness, compassion, courage, ability and discipline, e.g., being more or less developed, evolved and enlightened as a real and true human being. Some of our subordinate ego-selves may possess these qualities to a greater or lesser degree, e.g., more or less wise and compassionate.

5) Whether or not the organizing and regulating of our subordinate ego-selves is effectively accomplished by our executive ego-self of our whole ego-Self, e.g., feeling whole, fulfilled and comfortable or fragmented, dissatisfied and agitated. Some of our subordinate ego-selves may be more or less dominant, e.g., more or less assertive and aggressive.

6) Whether or not our subordinate ego-selves of our whole ego-Self are well-disciplined, compliant and strong, e.g., feeling or not feeling a sense of well-being, confidence and personal power. Some of our subordinate ego-selves may be more or less subordinate, e.g., more or less self-assured and agreeable.

By means of the above determinations, intrapersonal oppositions and conflicts and their nature and the likelihood of either their resolution or perpetuation can be predicted.

Our executive ego-self of our whole ego-Self will either be or not be able to effectively, successfully and usefully: 1) integrate and balance our various conflicting subordinate ego-selves and resolve conflicts between them; 2) maintain the unity of our whole ego-Self and 3) preserve the ultimacy, presence and radiance of True Tao-Nature.

In doing so, our effective and successful executive ego-self also considers generally advantageous situations and circumstances in intrapersonal conflict and employs and modifies strategic tactics that are specific to, and appropriate for, the particular subordinate ego-self that is being related to, e.g., quietly contemplating and reflecting on positive qualities of a subordinate ego-self or actively role-playing and dialoguing with its negative qualities.

Our subordinate ego-selves of our whole ego-Self can oppose, compete and conflict with each other for autonomy, independence, superiority and dominance and with our executive ego-self for sovereignty, hegemony, power and control.

The Tao/Way/Art of conflict and its resolution between our subordinate ego-selves and with our executive ego-self of our whole ego-Self can involve our executive ego-self deceiving or tricking our opposing subordinate egos in various ways in the interest of, and as a means to the end of, preserving the wholeness of our ego-Self, e.g., by strategically pretending to include and accept an undesirable trait as part of our otherwise positive self-image and self-concept.

Creating a sense of being and feeling our whole ego-Self with our harmoniously integrated executive ego-self and our subordinate ego-selves can be accomplished by our executive ego-self:

1) Enticing our subordinate ego-selves by pointing out advantages of being equally included.

2) Minimizing its executive status and then overcoming our opposing subordinate ego-selves.

3) Being amply prepared for dealing with our strong and opposing subordinate ego-selves.

4) Evading our strong and opposing subordinate ego-selves by ignoring and avoiding them.

5) Riling our angry opposing subordinate ego-selves by annoying, irritating, aggravating them.

6) Feigning weakness and provoking the arrogance of our opposing subordinate ego-selves.

7) Tiring our opposing subordinate ego-selves when they appear to be resting and inactive.

8) Separating and dividing our opposing ego-selves when they are collectively united.

9) Dealing with our opposing subordinate ego-selves when unexpecting and unprepared.

Our executive ego-self of our whole ego-Self doesn't let our opposing subordinate ego-selves know of plans, strategies and tactics in order to keep them in equally subordinate positions, e.g., by only concentrating and focusing upon itself, its intentions and objectives.

Our executive ego-self of our whole ego-Self that is effective in dealing with and successful in commanding our opposing and competing subordinate ego-selves and keeping them equally compliant; takes considerable preparatory time to deliberate, calculate and to make and practice plans, strategies and tactics before any direct engaging and interacting with them.

This is the Tao/Way/Art: 1) of enabling the foreseeing of success or failure in dealing with and resolving our intrapersonal or intrapsychic conflicts, 2) of increasing the likelihood of being able to effectively and successfully include, integrate, regulate and manage our subordinate ego-selves, 3) to create, maintain and sustain the integration of our whole ego-Self and 4) to insure and safeguard the radiant presence of our True Tao-Nature and Tao-State.

Going to war
is a decision to deplore
and an activity to abhor.

Your Reflections

CHAPTER TWO

WAGING WAR

作　　　　戰

TSO/ZUO
Make/Do

CHAN/ZHAN
War/Battle

JEN/REN – human being/person
+
CHA/ZHA – suddenly/at first

KO/KO – spear/lance
+
TAN/DAN – single/alone

THE TEXT

SUN TZU SAYS:

In the conduct and operations of war, when there are fielded thousands of chariots and hundreds of thousands of soldiers with provisions for a thousand miles, the expenditures are great; from smaller items like paint and glue to larger items like chariots, weapons and armor, and will reach costing thousands of ounces of gold or silver per day.

The purpose of engaging in war is to be victorious. When extended battles are engaged in; weapons will become dulled and soldiers demoralized. Laying siege to walled cities will exhaust their strength.

Protracted campaigns will deplete the State's resources and the cost of the military's supplies and operations will not equal the efforts involved and the losses undergone.

With dulled weapons, demoralized soldiers, exhausted strength and depleted resources and provisions; the enemy will take advantage of the extreme situation and even a wise General will not be able to avoid the ensuing negative consequences.

There has been foolish haste in ending wars but skillfulness has not characterized protracted ones and no country has ever benefited from them. Thoroughly understanding the drawbacks of war is the only way of understanding the reasons for the least costly ways of conducting it.

Troops of soldiers are not conscripted twice or supply-wagons loaded more than three times.

State funds become depleted and drained in order to maintain an army at a distance and maintaining an army closer causes high prices, impoverishes the people and dissipates most of their income. State expenses for broken chariots and worn-out horses, helmets and armor, spears and shields and oxen and wagons use up more than half of its revenue.

Foraging one quantity of food and supplies from the enemy saves twenty of one's own.

Defeating an enemy involves soldiers being angered and seeing defeating them as advantageous and an opportunity to obtain goods and to receive rewards for taking chariots, substituting war flags, using enemy chariots and treating captured enemy soldiers well. Doing so is using an enemy army to strengthen one's own.

Therefore, the object of warfare is a quick victory rather than a prolonged campaign. The General of an army is the decisive determining factor in the fate of soldiers, the people, the Ruler and the State; whether they all are safe and at peace or endangered and in peril.

THE SYNOPSIS

Waging war is very costly and depletes a country's resources and impoverishes its people.

In protracted wars, equipment and weapons degrade and soldiers' morale declines.

A country is thus vulnerable to the enemy and has never benefited from protracted wars.

Understanding the drawbacks of war is the only way of recognizing the need for economy. Maintaining an army far away or close depletes funds, raises prices and repairs are costly.

Food and supplies need to be foraged from the enemy and their chariots need to be taken.

Enemy soldiers need to be captured and their army must be used to strengthen one's own.

The object of war is a quick victory rather than a prolonged campaign in order to cut costs.

The General is the deciding factor in the safety and peace of a country, its Ruler and people.

THE COMMENTARY

Intrapersonal conflict and fighting inner wars within our whole ego-Self are costly for us to engage in, experience and deal with and they create dire conditions, situations and circumstances for us. They dispirit, distract and displace the unity, focus and concentration of our True Tao-Nature; deregulate, dissipate and deplete the resources, energy and strength of our whole ego-Self and disrupt, disorganize and deviate its integrity, functioning and activities.

Intrapersonal conflict, fighting and wars between our opposing subordinate ego-selves and between our opposing subordinate ego-selves and our executive ego-self fragment the healthy integrity and harmony of our whole ego-Self. Our conflicting and contending opposing subordinate ego-selves can actively, overtly and directly vie and fight: 1) with each other for hegemony, superiority, dominance and to be the strongest, greatest, best and finest and 2) with our executive ego-self to usurp and co-opt sovereignty, autonomy and authority and to be the highest, noblest, grandest and most supreme.

For example, all such intrapersonal conflicting and contending activities: 1) distract, take up and use up the psychic energy of our whole ego-Self, 2) misguide, deviate and preoccupy it with internal struggles and battles that enervate and weaken it and 3) render it unable to engage in and to carry on its usual functions, activities and relationships.

One or more of our subordinate ego-selves: 1) can become disgruntled with its lower status, requirements and duty to obey orders and commands from on high and to be strictly disciplined; 2) can be insubordinate, rebel, revolt and engineer insurrections and mutinies and 3) can attempt to prove itself worthy of replacing our executive ego-self and even to be the most able 'son of heaven' to assume, execute and fulfill the 'mandate of heaven' to gather, organize, command and lead our whole ego-Self and to rule and safeguard the Tao-State and our True Tao-Nature.

Having protracted relationships and dealings between our conflicting and warring opposing subordinate ego-selves over a lengthy period of time is debilitating, exhausting and disabling for our whole ego-Self and involve intensive inner power differentials and struggles and extensive inner weaknesses and vulnerabilities.

Proactively, quickly and skillfully dealing with our opposing and conflicting subordinate ego-selves benefits the overall solidity, stability, health and well-being of our whole ego-Self, e.g., by including them as equal constituents of our whole ego-Self that make essential, necessary and valuable contributions to it.

In order to maintain the overall integrity of our whole ego-Self, our opposing and conflicting subordinate ego-selves can also be disidentified from by our executive ego-self. Additional subordinate egos are not discovered and conscripted by our executive ego-self and energy resources are selectively used and are not continually made available, used, used up and need to be resupplied.

Our executive ego-self can economically conserve and not deplete, drain and waste its own vital energy and can obtain additional energy from our subordinate ego-selves themselves by, for example, using and converting the energy of their anger and other oppositional energies into advantageous opportunities for strengthening itself.

The object of intrapersonal conflicts for our executive ego-self *vis a vis* our opposing subordinate ego-selves is to resolve them quickly, swiftly and economically and to not waste time, energy and resources by engaging with them in prolonged baseless, senseless, pointless, valueless, meaningless and fruitless warring relationships.

Doing so, is a determining factor in protecting, preserving, maintaining and sustaining: 1) the ultimacy and sovereignty of the Tao-State, 2) the vitality and integrity of our True Tao-Nature, 3) the safety, security, strength and serenity of our whole ego-Self, 4) the solidity, stability and utility of our subordinate ego-selves and 5) the overall well-being, peacefulness, freedom and happiness of our human being.

For example, what if the effects and consequences of each confusing, disorganizing, imbalancing, destabilizing, divisive, alienating, stressful and upsetting intrapersonal conflict and each obsessively negative thought, each possessively attached feeling, each dysfunctionally maladaptive habit, each compulsively erroneous action and each oppositionally separated interrelationship reduced the longevity of our precious human life by one second, one minute, one hour, one day, one month or one year!

War is waged at high cost.
The lines of peace are crossed
and many lives are lost.

Your Reflections

CHAPTER THREE
STRATEGIC ATTACK

MOU/MOU
Plot/Stratagem

YEN/YEN – word/speak
+
MOU/MOU – a certain one

KUNG/GONG
Attack/Assault

P'U/PU – to tap/rap/knock
+
KUNG/GONG – art/work

THE TEXT

SUN TZU SAYS:

In the *Art of War*, it is optimal to keep one's country and army and its divisions whole and to capture the enemy's country and army and its divisions intact and to not destroy them. Fighting and conquering enemy armies in a hundred battles is not optimal excellence. Supreme excellence is strategically weakening and breaking the enemy's resistance and subduing them without engaging and fighting.

The highest form of a General's leadership is to block and thwart the enemy's strategic plans. The next best is to disrupt the enemy's alliances and avoid confrontation with enemy forces. The least best is to directly attack the enemy army on a battlefield. The worst is laying siege to fortified cities that requires heavy equipment and lengthy construction, involves disastrous outcomes and many soldiers dying and should be done only as a last resort. Quickly defeat and conquer enemy armies and states without fighting protracted battles and capture cities without attacking them. Weapons will not dull, morale will be preserved and advantages will be complete.

Rules in war are:

1) if forces are ten to the enemy's one, surround them; 2) if five to one, attack them; 3) if two to one, divide them; 4) if equally matched, engage them; 5) if fewer in number, avoid them; 6) if unequal in every way, retreat from them and 7) a smaller and weaker army, even when determined to fight, will still be captured by a larger and stronger one.

The General is the solid protective and defensive asset of the State. When protective defense is solid, the State remains strong and safe but, if defense is deficient or defective, the State becomes weak and vulnerable.

Three ways that a Ruler can create misfortune for the army are:

1) Crippling an army by ordering it to advance or retreat without knowing that it is not in a position to do so,

2) Not understanding military matters, interfering with an army as if administering a State and thus confusing officers and

3) Not understanding the balance of power in the army, utilizing army officers indiscriminately and thus creating doubt and suspicion among them.

These three ways hamper the effectiveness and successful-ness of the army by creating confusion, doubt, suspicion, unrest and a lack of confidence among soldiers and officers; making the army chaotic and vulnerable to being defeated by the enemy and thus throwing victory away.

The five essentials for victory are: 1) Knowing when and when not to fight, 2) Knowing how and when to use superior and inferior military forces, 3) Uniting and animating an army with the same spirit of victory throughout all of its ranks (*esprit de corps*), 4) Being prepared against an unprepared enemy army and 5) Having competent military leadership that is not interfered with by the Ruler.

Knowing the enemy and knowing oneself, victory results in a hundred battles. Knowing oneself but not the enemy, results in an equal number of victories and defeats. Knowing neither oneself nor the enemy, results in defeat in every battle.

THE SYNOPSIS

In the *Art of War*, it is optimal to keep one's country and army and those of the enemy whole.

Fighting and conquering an enemy is not excellence but subduing them without fighting is.

The best leadership is thwarting the enemy's plans and the next best is disrupting alliances.

The least best is fighting the enemy on a battlefield and the worst is laying siege to a city.

Quickly defeating and conquering enemy armies without fighting protracted battle is ideal.

If one's forces are ten to the enemy's one, surround them; if five to one, attack them.

If one's forces are two to the enemy's one, divide them; if one-to-one, engage them.

If one's forces are fewer in number, avoid them; if unequal in every way, retreat from them.

Even a very determined but smaller and weaker army will be defeated by a larger army.

The General is the solid protective defensive asset of a State and its safety and strength.

A Ruler can cripple an army by faulty orders, confusing interference and disrupting ranks.

Essentials for victory are: 1) knowing when and when not to fight; 2) how to use military forces; 3) uniting the whole army with the same spirit of victory throughout its ranks; 4) being prepared against an unprepared enemy and 5) not having the Ruler interfere with military leadership.

Knowing the enemy and oneself results in victory all of the time. Knowing oneself but not the enemy results in victory half of the time and not knowing either results in defeat all of the time.

THE COMMENTARY

Intrapersonal conflict is between our executive ego-self and our opposing subordinate ego-selves of our whole ego-Self and between our various opposing subordinate ego-selves. Chapter One of *The Art of War* focuses on: 1) the nature, structure and functions of our executive ego-self and our subordinate ego-selves, 2) the nature of and factors in intrapersonal relationships between our executive ego-self and our subordinate ego-selves, 3) preparations and assessments made by our executive ego-self for dealing with intrapersonal conflicts with our opposing subordinate ego-selves and 4) strategic tactics used by our executive ego-self in dealing with and resolving conflicts with and between our opposing subordinate go-selves.

Chapter Two of *The Art of War* focuses on: 1) the high costs and negative effects of intrapersonal conflicts and the integral value and necessity of quickly resolving them, 2) the nature and motivations of our opposing and conflicting subordinate ego-selves and 3) ways that our executive ego-self can advantageously use the energy of intrapersonal conflict to strengthen our whole ego-Self.

This Chapter Three of *The Art of War* focuses on: 1) the conduct of, and the reasons for, intrapersonal conflict resolution, 2) strategic tactics of successful conflict resolution, 3) the ways successful conflict resolution can be interfered with and 4) essential necessities of successful conflict resolution.

In general, in intrapersonal conflicts, our executive ego-self of our whole ego-Self deals with our opposing and conflicting subordinate ego-selves through: 1) bringing them into its field of conscious awareness through imaging and visualizing, 2) establishing and solidifying a connection with them in consciousness, 3) understanding and contemplating their uniquely individual nature, qualities and characteristics, 4) inviting them and asking them to share what their particular interests, desires and motives are for opposing and conflicting and 5) engaging in an active dialoguing and role-playing with them as to possible alternatives to, compromises in and resolutions of conflicts.

In the Tao/Way/Art of intrapersonal conflict, it is important and optimal to maintain the wholeness and strength of the Tao-State, our True Tao-Nature, our whole ego-Self, our executive ego-self and our subordinate ego-selves without dividing, separating or fragmenting them, e.g., our executive ego-self holds all of our subordinate ego-selves equally in its integral consciousness.

Overcoming our opposing and conflicting subordinate ego-selves a hundred times is not as excellent as strategically weakening their resistance and gaining the advantage without fighting with them, e.g., our executive ego-self does not achieve the harmonious inclusion, organization and cooperation of our subordinate ego-selves by defeating, conquering and subduing them through aggressively forceful actions.

The highest form of our wise, effective and successful executive ego-self's leadership is to block, thwart and disarm the intentions, plans, coping strategies and tactical maneuvers of our opposing and conflicting subordinate ego-selves, e.g., our executive ego-self achieves the harmonious inclusion, organization and cooperating of our subordinate ego-selves by frustrating their actions with baffling counteractions.

The next best is to disrupt any of their alliances and to avoid direct engagement with them, e.g., our executive ego-self maintains the separateness of our collectively opposing subordinate ego-selves within the field of conscious awareness and thus weakens their power.

The least best is to directly engage them in conflict in the field of conscious awareness of our whole ego-Self, e.g., when absolutely unavoidable, our executive ego-self uses direct strategic tactics to subdue, control and organize our opposing subordinate ego-selves into compliance.

The worst, as a last resort, is to persistently attack and lay siege to their strong walled-up ego-defenses until they are ineffective, which results in our subordinate ego-selves being useless, e.g., as a last resort, our executive ego-self disarms the ego-defenses of our opposing and conflicting subordinate ego-selves even though this renders them useless for our whole ego-Self in conflict resolution.

Our wise, effective and successful executive ego-self deals with our opposing and conflicting subordinate ego-selves early and as quickly as possible and acts only when it is advantageous and beneficial to do so and without engaging in protracted conflicts with them or making prolonged attacks on their ego-defenses, e.g., our skillful executive ego-self: 1) prevents conflicts with our opposing subordinate ego-selves from occurring by anticipating their possible actions early, 2) intervenes with them only when that serves our whole ego-Self and 3) does not waste time and energy conflicting with them in prolonged engagements.

General rules that our wise, effective and successful executive ego-self follows in dealing with and resolving intrapersonal conflicts with our opposing subordinate ego-selves are:

1) Surround them if the ego-strength of our executive ego-self is 10 times theirs, e.g., our opposing subordinate egos can be induced to yield to the strength of our whole ego-Self and to be included within its encompassing orb of consciousness.

2) Attack them if the ego-strength of our executive ego-self is 5 times theirs, e.g., our executive ego-self only engages in direct confrontation with our opposing subordinate ego-selves when able to be successful.

3) Divide them if the ego-strength of our executive ego-self is

2 times theirs, e.g., our executive ego-self strategically and tactically separates our fairly strong opposing subordinate ego-selves within its conscious awareness to divide their power, weaken them and facilitate dealing with them.

4) Engage them if the ego-strengths of our executive ego-self and our subordinate ego-selves are equal, e.g., when the strengths and power of our executive ego-self and our opposing subordinate ego-selves are equal, our executive ego-self directly engages them in intrapersonal conflict within its conscious awareness.

5) Avoid them if the ego-strength of our executive-ego is less than theirs, e.g., when the strength of our opposing subordinate ego-selves exceeds that of our executive ego-self, it minimizes their power by temporarily putting them aside and postponing any direct engagement with them within its conscious awareness.

6) Retreat from them if the ego-strengths of our executive ego-self and our opposing subordinate ego-selves are completely unequal, e.g., when the strengths and power of our opposing subordinate ego-selves are overwhelming, our executive ego-self backs off, disidentifies from and rebuilds and reinforces its own strengths within its conscious awareness.

7) A few of our opposing and conflicting subordinate ego-selves, even if very determined to fight, will still be overcome by our stronger executive ego-self, e.g., the strength of our executive ego-self and its strategic energy potential generally sufficiently exceed those of our opposing subordinate ego-selves even when they are committed and determined to win a conflict.

Our executive ego-self is the solid defender and stable protector of the Tao-State, our True Tao-Nature and our whole ego-Self. When protective defense is complete, the Tao-State and our whole ego-Self are strong and not weak, safe and not vulnerable, e.g., it is the completeness, solidity, stability and strength of our enlightened, wise, compassionate and skillful executive ego-self that maintains and sustains the viability, integrity and vitality and the safety, stability and strength of the Tao-State, our True

Tao-Nature and our whole ego-Self.

The ways that our obscured, obstructed and displaced True Tao-Nature can create misfortune for our whole ego-Self in intrapersonal relationships and conflict are:

1) Crippling it by not discriminating when it is and is not in a position to advance or retreat, e.g., not being aware of the actual realities of our whole ego-Self and its many ego-selves.

2) Interfering with it by not clearly discriminating between realities of conflict and peace, e.g., not making whole ego-Self integral and form differentiations in conscious experience.

3) Disrupting it by not discriminating in the balanced organization of the chain of command, e.g., not making distinctions as to the functional organization of our subordinate ego-selves.

These three ways hinder our whole ego-Self of by: 1) creating distrust, confusion, doubt, insecurity, uneasiness and chaos among our executive ego-self and our subordinate ego-selves, 2) creating internal vulnerability of our executive ego-self to our opposing subordinate ego-selves and 3) creating and perpetuating intrapersonal conflicts and derailing chances for our executive ego-self to effectively deal with and to successfully resolve them.

The five essentials necessary for our executive ego-self to effectively deal with and to successfully resolve intrapersonal conflicts with our opposing subordinate ego-selves are:

1) Knowing when and not when to engage and contend with them, e.g., our executive ego-self chooses appropriate times when it has a strong presence within its conscious awareness.

2) Knowing how and when to use and not use superior and inferior skills, e.g., our executive ego-self is competent and proficient in using its strong capabilities and appropriate abilities.

3) Uniting and animating all of them with the same one spirit throughout, e.g., our executive ego-self is able to animate one *esprit de corps* among our conflicting subordinate ego-selves.

4) Being prepared for and against them with strategic and tactical plans, e.g., our executive ego-self anticipates potential

conflicts and plans appropriate strategic tactical maneuvers.

5) Exercising capable leadership and decision making without interference, e.g., our executive ego-self makes independent, swift, decisive moves that end conflict quickly.

Effectively dealing with conflicts, overcoming our opposing and conflicting subordinate ego-selves and successfully resolving conflicts by our executive ego-self requires knowing our opposing subordinate ego-selves, itself and our whole ego-Self, e.g., our executive ego-self needs to thoroughly understand and to be in touch with our various opposing subordinate ego-selves, its own self and our whole ego-Self and with the nature, kinds and dynamics of their intrapersonal relationships with each other.

Knowing itself and our whole ego-Self but not knowing our opposing and conflicting subordinate ego-selves results in overcoming them half the time, e.g., our executive ego-self needs to thoroughly understand our various opposing subordinate ego-selves and their unique natures, intentions, strengths and weakness and strategic tactics and maneuvers, and not just itself and our whole ego-Self, in order to wholeheartedly deal with and to fully resolve intrapersonal conflicts.

Not knowing itself, our whole ego-Self and our opposing and conflicting subordinate ego-selves results is not overcoming them, e.g., our executive ego-self when not thoroughly understanding our opposing subordinate ego-selves, its own self and our whole ego-Self is unable to effectively deal with and to successfully resolve intrapersonal conflicts.

In order to create, maintain and sustain the integrity, solidity and stability of our whole ego-Self; our executive ego-self needs to be aware of, accepting of, connected with and understanding of our various subordinate ego-selves and to fully integrate, continuously support, clearly command and consistently discipline them, e.g., in our undivided, integrated, unconflicted and whole ego-Self; our executive ego-self and our various subordinate ego-selves are intrapersonally in constant, flexible, fluid

and ever-changing positive contact, relationship, interaction
and communication within our conscious awareness.

By strategy, enemies are defeated,
certain victories are greeted
and resources undepleted.

Your Reflections

Chapter Four
Military Forms

軍　形

Chun/Jun
Army/Military

Ch'e/Che – a cart/carriage with axle, two wheels, body/ to roll/crush

+

Mien/Mian – roof/shelter

Hsing/Xing
Form/Figure

Shan/Shan – feathers

+

Ch'ien/Qian – two shields

The Text

Sun Tzu says:

Good fighters skillfully place themselves beyond the possibility of being defeated and wait for vulnerable opportunities to defeat the enemy. Securing against defeat is one's doing and defeating the enemy relates to their doing. The good fighter is able to be secure against defeat but cannot cause the vulnerability of the enemy. Thus, one may know how to defeat and conquer an enemy but may not be able to do it.

Security against defeat involves defensive strategies and tactics and ability to defeat the enemy involves taking the offensive. The defensive position is taken when strength is weak and the offensive position is taken when strength is strong. The defensive General protectively hides in the deepest recesses of the earth and the offensive one victoriously springs forth from the highest acme of heaven.

Seeing victory that is only popularly known and praised by the masses is not excellence. Lifting a hair is not great strength. Seeing the sun and the moon and hearing the sound of thunder are not indications of keen vision and acute hearing.

What characterizes an excellent fighter is not only winning battles but winning them with ease and without seeking or gaining a reputation for wisdom or credit for bravery. Battles are won by not making errors and by establishing the certainty of victory which is conquering an enemy that is already defeated.

The excellent and skillful fighter adopts a position which makes defeat impossible, assures victory and does not miss any opportunities for defeating the enemy. In war, the victor only seeks a battle after the victory has already been won and the vanquished initially battles and then seeks victory.

The excellent Commanding General cultivates Tao/The Way and strictly adheres to military rules, strategic method and troop discipline and therefore has the power to control, influence and determine successful outcomes.

The strategic art of military method involves measuring, estimating, calculating, weighing probabilities and gauging ultimate victory. Measurement depends upon terrain, estimation upon measurement, calculation upon estimation, weighing probabilities upon calculation and gauging ultimate victory upon weighing probabilities.

A victorious army compared with a vanquished one is like comparing a heavy weight to a single rice-grain. The onrushing advance of a conquering military force is like the bursting forth of dammed up waters into a thousand foot deep chasm. This is the strategic formation, positioning and disposition of the army.

THE SYNOPSIS

Good fighters place themselves beyond being defeated and wait to defeat the enemy.

Securing against defeat is one's own doing and being defeated is the enemy's own doing.

Securing against defeat uses defensive tactics, defeating the enemy uses offensive ones.

The defensive position is taken when the army is weak and the offensive one when strong.

Victories praised by the masses are not excellence. Excellence is winning the wars easily.

Battles are won by not making errors and conquering an enemy that is already defeated.

An excellent fighter makes defeat impossible and misses no chances to defeat the enemy.

The victor battles after the victory is won. The vanquished battles and then seeks victory.

An excellent General cultivates Tao and strictly adheres to military rules, method, discipline.

The art of military strategy involves measurement, estimation, calculation, probabilities.

A victorious army is a heavy weight against a rice-grain, the onrush of dammed up water.

This is the strategic formation, tactical positioning and disposition of the armed forces.

THE COMMENTARY

Our wise, effective and successful executive ego-self of our whole ego-Self skillfully places itself in an unassailable position that is beyond the possibility of being defeated by our opposing subordinate ego-selves and awaits their advance and the opportunity to exploit their vulnerabilities and to defeat them, e.g., the internal safety, security, solidity and stability of our whole ego-Self depends upon the control of our opposing subordinate ego-selves by our executive ego-self that places itself in a strong, elevated and well-fortified position so as not to be defeated by them and so as to enable defeat of them.

Securing itself and our whole ego-Self and not creating opportunities for our conflicting subordinate ego-selves is the doing of our executive ego-self but being able to defeat them is their own doing, e.g. our overzealous conflicting subordinate ego-selves can defeat themselves by their fervor to attain supremacy.

Our wise, effective and successful executive ego-self is able to be secure against being defeated by our conflicting subordinate ego-selves, but is unable to cause their vulnerability which, again, is their doing, e.g., our executive ego-self does not have complete control over the strengths and weakness of our conflicting subordinate ego-selves. Therefore, our executive ego-self may know how to defeat our conflicting subordinate ego-selves but does not have the necessary control of them in order to do so.

Security against being defeated by our conflicting subordinate ego-selves involves defensive strategic tactics but defeating them requires taking the offensive. A defensive position is taken when our executive ego-self is weaker than our conflicting subordinate ego-selves and an offensive one is taken when the reverse is the case, e.g., our executive ego-self takes the offensive only when it is stronger than our opposing subordinate ego-selves.

The defensive position is being protectively concealed within the deepest recesses of the groundedness of our whole ego-Self and the offensive position is springing forth revealed from the highest expanses of the spaciousness of our whole ego-Self, e.g., our executive ego-self stands in its solid lower connection with earth and moves from its higher spacious position of heaven .

When our executive ego-self only considers defeating our conflicting subordinate ego-selves from the ordinary viewpoint of the populous, it is no real victory. Lifting a hair is not real strength. Seeing the sunlight is not keen vision. Hearing a thunderclap is not acute hearing, e.g., our executive ego-self's effective and successful use of the power of its more grounded and spacious positions is natural and just a part of our whole ego-Self and our True Tao-Nature.

What characterizes the effectiveness and successfulness of our wise and courageous executive ego-self in dealing with and resolving intrapersonal conflict is not just overcoming and defeating a large number of our conflicting subordinate ego-selves, but is doing so easily and quickly and without seeking or obtaining a reputation for wisdom or bravery, e.g., successes are not ego-invested attainments and are due to not making errors in struggles and conflicts and by establishing the certainty of victory by overcoming our opposing subordinate ego-selves that are already defeated before they begin conflict.

Our wise, skillful, effective and successful executive ego-self assumes a position that is undefeatable and does not miss any advantageous opportunities to defeat our conflicting subordinate ego-selves, e.g., the grounded and spacious positions of our executive ego-self are deeply beneath and vastly beyond the reaches of our conflicting ego-selves and create advantages and opportunities to engage and defeat them if and when it is necessary to do so.

In intrapersonal conflicts, our victorious executive ego-self only engages in a battle with our opposing subordinate ego-selves after a victory over them has already been won. Our vanquished opposing subordinate ego-selves begin with a battle with our executive ego-self and then seek a victory over it, e.g., our executive ego-self has already defeated our conflicting subordinate ego-selves through indirect means that do not require direct engagement in fighting with them.

Our excellent executive ego-self cultivates and accords with Tao/The Way and adheres to martial law and military rules, effective and successful strategic tactics and the disciplining of our subordinate ego-selves and, by doing so, has the power to control, influence and determine positive outcomes in intrapersonal conflicts with them, e.g., our excellent executive ego-self is in enlightened accord with the ultimate reality, absolute and relative law and method of Tao and the intimate energy and virtuosity of its individualized power/Te.

The Tao/Way/Art of employing strategic tactics in conflicts with our opposing subordinate ego-selves involves measuring, assessing, calculating and weighing probabilities of their advantageous, effective and successful use, e.g., our executive ego-self carefully considers and utilizes the most optimal means and methods for ensuring control over our subordinate ego-selves and the viability, integrity and vitality of our whole ego-Self.

Our victorious whole integral ego-Self as compared with a vanquished one, is like a heavy weight compared to a rice-grain. The on-rushing advance of our conquering whole energized ego-Self is like the bursting forth and momentum of dammed up waters plummeting into a thousand foot deep chasm.

Such are: 1) the synergic power of our wise, compassionate, skillful, effective, successful and excellent executive ego-self and our conflict-free and harmonized subordinate ego-selves and 2) their strong defensive, offensive and winning strategic tactical formations, positionings, dispositions and maneuvering that safeguard, preserve and sustain the peace of the Tao-State, our True Tao-Nature and our whole ego-Selves.

CLOSING COMMENT

This concludes Chapters One, Two, Three and Four on intrapersonal relationships and conflict. Intrapersonal conflict is successfully resolved by, through, in and as the harmonious co-existing and cooperative interrelationship of our executive ego-self with our numerous complementary and interdependent subordinate ego-selves within our whole ego-Self. As such, this preserves the original unity, integrity, peacefulness and pacifism of our True Tao-Nature and Tao-State within our awakened, enlightened and liberated consciousness, conscious awareness and conscious experience of our intrapersonal relationships.

Before starting war, be pensive.
Better to take the defensive
rather than the offensive.

The place to begin
is the conflict within.
Effecting its resolution
is the ego-self's solution,
higher Self's integration,
True Nature's creation.

Some Examples

Since the commentaries are generalized abstractions of intra-personal relationships, it may be helpful to provide some concrete examples of the relationships between our executive ego-self/I, subordinate ego-selves/me-s and whole ego-Self/human being.

1) Our executive ego-self/I may identify with and use its challenging sub-personality/me to deal with an opposing rebellious one/me that is attempting to usurp its authority and would disrupt the harmonious integrity of our whole ego-Self/human being.

2) Our executive ego-self/I may mediate and resolve an internal conflict between a mild sub-personality/me and a wild sub-personality/me in order to maintain the harmonious integration of the whole ego-Self/human being.

YOUR REFLECTIONS

INTERPERSONAL RELATIONSHIPS

WARRIORS

CHAPTER FIVE
SOLDIER POWER

兵　　　勢

PING/BING
Soldier/Military

SHIH/SHI
Power/Momentum

PA/BA – eight/to divide
+
CHIN/JIN – a battle ax

LI/LI – strength/force
+
CHIH/ZHI – hold/direct

THE TEXT

SUN TZU SAYS:

The commanding, controlling and fighting of a large military force is the same as that of a small one by virtue of their forming, organizing, positioning and communicating. To withstand an enemy attack involves maneuvers that are both overt, direct, orthodox, conventional, planned, fixed and known and covert, indirect, unorthodox, unconventional, spontaneous, flexible and surprising.[6]

The force of a strong army's impact on an opposing weaker one is like the power of a heavy stone thrown against an egg and involves assessing an opposing army's weak and strong points, its emptiness and fullness.

In any fighting, direct maneuvers are used in engaged battles and indirect maneuvers are used to gain victories. Efficiently implemented indirect, spontaneous and surprise strategic tactics are as unlimited as wide and spacious heaven-earth, unending as flowing riverstreams and as continually recurrent as the sun-moon and the four seasons.

There are only five musical notes, primary colors and cardinal tastes; yet in combination they give rise to more melodies, hues and flavors that can ever be heard, seen and tasted.

Likewise, in battles, there are only two methods of attack, the direct and the indirect, but which in combination, give rise to endless strategic tactical maneuvers. Both methods produce each other and move in an endless circle of inexhaustible possibilities. How can they be exhausted?

The energy of onrushing military troops is like the momentum of a raging flood moving heavy stones along its course. The quality of military decisions is like the perfectly timed precise swooping of a falcon catching its prey.

So likewise, the energy of good fighters is strategically configured, tactically directed, gains momentum and is precise, forceful, overpowering and successful. The advantageous potential of their energy is like the cocking of a crossbow and the timed precision of their decisions is like its triggered release.

During the turmoil and chaos of battles, military troops may appear disordered and confused, yet they will still be undefeated. Disorder, fear and weakness entail order, courage and strength based upon organization, potential energy and strategic tactical formation and maneuvering.

Skillfully effective and successful Generals keep enemy armies moving via shifting military formations. Unimportant things are sacrificed and bait is held out to keep enemy armies moving into a position where they can be ambushed by the full force of military power.

Skillfully effective and successful Generals relate to the army of military troops as one whole, do not require excessive things from individual soldiers, pick elite soldiers for front lines and use their combined energies, power and influence.

As a result, the ensuing strategic tactical advantage and potential power of soldiers are like round boulders and logs that remain still on level ground but easily roll down slopes and rapidly gain momentum. The intensity, momentum and influence of the increasingly powerful energy of soldiers are like round boulders and large logs rolling down a thousand foot high mountain.

Such is the matter of the strategic potential energy, strategic tactical configuration and the advantage and disposition of military soldier power.

The Synopsis

The commanding, controlling and fighting of a large and a small military force is the same.

Withstanding an enemy attack requires both indirect and direct methods and maneuvers.[6]

Weak and strong points, emptiness and fullness determine indirect and direct maneuvers.

Direct maneuvers are used in engaged battles and indirect surprise methods gain victories.

Direct and indirect methods and maneuvers of attack used in concert are inexhaustible.

The energy of an onrushing army is like a raging flood, its precision like a swooping falcon.

Fighters' energy is strategically configured, gains momentum, is precise and overpowering.

Fighters' energy is a cocked crossbow and their precise decisions like its triggered release.

Skillful Generals keep armies moving and offer the enemy bait to move them into ambushes.

Effective Generals relate to the army as one whole, pick elite soldiers and use their power.

The powerful momentum of soldiers is like round boulders rolling down high mountains.

This is the matter of the energy and power of soldiers and its advantages and disposition.

THE COMMENTARY

Interpersonal conflicts are those occurring between our whole ego-Self and/or our subordinate ego-selves and those of an opposing whole ego-Self and/or its subordinate ego-selves.

Only a solid, harmoniously integrated and stable whole ego-Self that has effectively dealt with and has successfully resolved its intrapersonal conflicts with and between its executive ego-self and subordinate ego-selves can effectively, successfully and completely resolve interpersonal conflicts with an opposing whole ego-Self, e.g., ongoing unresolved intrapersonal conflicts within the whole ego-Self take focus and energy away from its availability, and weaken its ability, to deal with and resolve conflicts with an opposing whole ego-Self.

Effectively dealing with and successfully resolving conflicts with an opposing whole ego-Self is best accomplished when our whole ego-Self is a harmoniously and peacefully integrated unit with no active and compelling concurrent intrapersonal conflicts between our executive ego-self and our subordinate ego-selves that distract concentration, divide focus, misdirect effort and dissipate energy, e.g., the energy and power of the complete integration of our executive ego-self and our subordinate ego-selves of our whole ego-Self make it unassailable, impervious to attack and invulnerable to harm and defeat.

Our executive ego-self handles interpersonal conflicts between our whole ego-Self and an opposing whole ego-Self in many of the same ways that strategies and tactics are used in dealing with and resolving intrapersonal ones with our opposing subordinate ego-selves. In certain requisite situations, our executive ego-self can be completely identified with and as one of our subordinate ego-selves that is appropriately utilized to effectively deal with, and to successfully resolve, conflicts with an opposing whole ego-Self and to thus preserve the integrity of our whole ego-Self.

In confronting, effectively dealing and successfully resolving conflicts with an opposing whole ego-Self, both direct and indirect strategic tactics are used, e.g., planned polite apologies and agreement and spontaneously abrupt digs and surprises.[6]

Strong or full characteristics and empty or weak ones of an opposing and conflicting whole ego-Self are assessed, e.g., is an opposing and conflicting whole ego-Self more or less consciously aware, in touch with itself, energized, quick-witted, able to confidently support and clearly express itself, etc..

The power, strength and force of our whole ego-Self against the weakness of an opposing whole ego-Self can be like hurling a heavy stone against an egg. The power balance between our whole ego-Self and that of an opposing whole ego-Self is upset by surprise moves that catch it unawares and off guard and move and place it in disadvantageous and vulnerable positions, e.g., strong and unpredictable surprise strategic tactical maneuvers are ones that an opposing whole ego-Self doesn't expect, is unprepared for and defenseless against.

Direct and indirect strategies, tactics and maneuvers are complementary; mutually support, balance and augment each other and, when creatively employed in various combinations, are powerful, intense and inexhaustible, e.g., being both planned, thoughtful, understandable and limited and unplanned, spontaneous, unfathomable and limitless.

The powerfully advancing and onrushing energy of our harmoniously and fully integrated whole ego-Self is like the unstoppable momentum of a raging flood moving heavy boulders along its way, e.g., naturally, forcefully, seamlessly, frictionlessly and effortlessly.

Strategies, tactics and maneuvers used by our whole ego-self to confront, deal with, manage and resolve conflicts with an opposing whole ego-Self are accurately timed like the precision of a swooping hawk seizing its prey, e.g., finding the exactly right moment to intervene, make a point or contravene.

Likewise; the energy, power and force of our effective and successful whole ego-Self is strategically and tactically configured, precisely timed and promptly released and gains a momentum that completely overwhelms an opposing and conflicting whole ego-Self, e.g., when interpersonal confrontations are backed by and made with a high energy charge.

The strategic tactical advantage of the potential energy of our whole ego-Self is like the cocking of a crossbow and its implementation is like releasing its trigger, e.g., concentrating and building up the inner strengths and energy of our whole ego-Self before fully drawing upon and forcefully expressing them in an interpersonal conflict.

During intense confrontations, deep disagreements and heated arguments; our whole ego-Self may, at times, be or appear confused, disordered, uncontrolled, weak or acquiescent but still can be undefeated, e.g., managing these conditions by taking a time out break to recover the basic clarity, order, control, energy, strength and assertiveness of our whole ego-Self.

Our skillful, effective and successful whole ego-Self keeps an opposing and conflicting whole ego-Self engaged and constantly moving, e.g., by quickly responding and interrupting that causes and requires immediate reactive or responsive counter-moves.

Our skillful, effective and successful whole ego-Self ignores unimportant issues brought up by an opposing and conflicting whole ego-Self and draws it into engagement with baits, e.g., by agreeing with an opponent and inducing their false sense of superiority and confidence.

This maneuvers an opposing whole ego-Self into positions where it can be ambushed by the waiting full energy and force of our whole ego-Self, e.g., consciously leading or allowing an opposing and conflicting whole ego-Self to proceed along its path while building up energy for a well-timed winning strategic tactical move.

When being involved in interpersonal conflict, our skillful, effective and successful whole ego-Self 1) suspends any other concerns, 2) focuses only on the matter at hand, 3) maintains the wholeness of our executive and subordinate ego-selves and 4) engages an opposing whole ego-Self with those of our subordinate ego-selves that are the strongest and most relevant and useful, e.g., selecting and using aspects of our whole ego-Self that match and/or exceed those of an opposing whole ego-Self.

The result of doing so creates a strategic tactical advantage and overwhelming force that is like that of a heavy round boulder gaining unstoppable momentum as it rolls down a thousand foot high mountain.

Such is the effective strategic configuration and tactical advantage and the successful positioning and disposition of the powerful energy of our harmoniously integrated whole ego-Self and subordinate ego-selves actively, directly and fully engaged in interpersonal conflicts with an opposing and conflicting whole ego-Self.

> Soldier energy and power
> make the enemy cower
> and victories to flower.

Your Reflections

CHAPTER SIX
WEAKNESS AND SOLIDITY

虛 實

HSU/XU
Weak/Empty

SHIH/SHI
Solid/Substantial

HU/HU – tiger/bravery
+
HSU/XU – unreal/untrue/
humble/modest

MIEN/MIAN – a roof/shelter
+
KUAN/GUAN – real wealth/
cash strung together

THE TEXT

SUN TZU SAYS:

An army that is first in the battlefield and awaiting an enemy army is relaxed and fresh and an army that is second has to hurry to battle and arrives fatigued. Astute Generals compel enemy armies to approach their own armies and not vice versa. They strategically extend advantages to an enemy army and/or inflict minor damage so as to entice its approach through the prospect of gain and/or the fear of harm.

Relaxed enemy armies can be harassed, well supplied enemy armies can be starved and encamped enemy armies can be forced to move. One's army can appear where an enemy army must hurry to defend and can swiftly move to places that are unexpected by the enemy army.

When one's army can march through land where no enemy army is, they can do so without distress. Successful attacks on an enemy army can be made in its undefended places. The defensive safety of one's army can be ensured by holding positions that cannot be attacked by the enemy.

The successful General is skillful in offensively attacking an enemy army that doesn't know what, where, when or how to defend itself against and in defensively protecting what, where, when and how the enemy doesn't know to attack. This is the divine art of mystery, subtlety and secrecy; whereby one's army is formless, invisible and inaudible and an enemy army's fate is held in its hand.

Irresistible advances can be made when the weak points of an enemy army are attacked. Safe evasion of the advances and pursuits of an enemy army can be made if swift elusive movements are quicker than theirs.

When deciding to fight, even though an enemy army is defensively sheltered behind high ramparts or within deep trenches, it can be forced into combat by attacking some other place to which it will move to rescue. If deciding not to fight, the movements of an enemy army can be thwarted and it can be prevented from fighting by throwing it off course.

By staying invisible and knowing an enemy's visible dispositions, one's armed forces can be concentrated to form a single united army while dividing the enemy's army. This makes one's army whole and many to the enemy's separate and few, outnumbers it and enables attacking their weaker and inferior forces with one's own stronger and superior ones.

The place where fighting is intended is kept secret and an enemy's army is made to prepare for an attack at any possible place and required to distribute their forces in multiple places and thereby lessen their numbers and strength. If the enemy strengthens its vanguard and the left, it will weaken its rear guard and the right and vice versa. Having to prepare everywhere will weaken the whole enemy army even more. Numerical weakness comes from requiring an enemy army to prepare for possible attacks anywhere and everywhere and numerical strength comes from forcing them to do so.

Knowing the place and time of an upcoming battle, one's army can concentrate itself at the best distances to fight. Otherwise, vanguard and rear guard and right and left wings will be unable to support each other, especially when they are farther apart.

Though the soldiers of an enemy army may be greater in number, it doesn't mean that they will be victorious. They can be prevented from fighting by dispersing their numbers and by discovering their plans and the likelihood of their success. They can be stirred up in order to learn about their activity. They can be forced to reveal themselves and disclose weak and vulnerable places. It is important to compare armies as to their relative strengths and weaknesses.

When making strategic tactical formations, positions, dispositions and maneuvers, it is critical that one's army be formless with no fixed ones and to conceal them in order to be protected from the wisest Generals and subtlest spies. It is necessary to understand how victory can be made and had from the enemy army's own tactics and to 'turn the tables' on them.

The overt strategic tactics of an army that lead to victory can be seen but the covert ones by which victory is ensured cannot. Tactics used on one occasion to attain victory should not necessarily be repeated and need to be flexibly adjusted to new and ever-changing circumstances.

Effective military strategic tactics are like the natural course of water, running down from high places into low places. In warfare, this means strategically and tactically avoiding what is full and strong and attacking what is empty and weak. The shape and flow of water accords with the nature of its ground and successful military strategies and tactics are worked out in

accord with the enemy's formations.

Just as water has no constant shape, in warfare there are no constant conditions, formations, positions and dispositions. Generals who are able to modify tactics in relation to a particular unique enemy and thus defeat them show heaven-given divine and spiritual military genius.

The five elements of water, fire, wood, metal and earth are not equally dominant. The four seasons constantly, continuously and continually follow each other. Daylight comes and goes. Moonlight waxes and wanes. Everything changes.

THE SYNOPSIS

Armies arriving to a battle first are fresh, ones arriving second hurry and are exhausted.

Astute Generals compel an enemy army to approach their army first and not vice versa.

They do so by extending advantages or creating minor damage to entice or frighten them.

Relaxed armies are harassed, supplied ones starved, encamped ones are forced to move.

One's army can appear where an enemy has to hurry to defend and in unexpected places.

The successful General defensively protects what and where an enemy is unable to attack.

The successful General offensively attacks what and where an enemy is unable to defend.

Effective strategies are attacking weaknesses of an enemy army and by swift elusive moves.

Effective strategies are attacking places to which an enemy army needs to rush to defend.

Effective strategies are remaining invisible and united and dividing the enemy's army.

Effective strategies are keeping a battleground secret and

forcing an enemy to fortify many places at once, thereby decreasing their numbers and weakening their positions.

Numerical strength comes from forcing an enemy army to prepare for possible attacks.

Knowing the place and time of a battle, one's army can be at the best distance to fight.

Smaller armies can discover plans of larger ones and stir them up to observe activities.

Smaller armies can force larger ones to reveal their position and any vulnerable places.

Tactical formations and dispositions must be formless, unfixed and protected from spies.

It can be understood how victory can be accomplished by using the enemy's own tactics.

Overt tactics leading to victory can be seen but covert tactics are ones that insure victory.

Successful tactics are not repeated and need to be adjusted to new, changing situations.

What is full and strong is tactically avoided and what is empty and weak is attacked.

There are no constant or fixed conditions in warfare, positions, formations, dispositions.

A General who can flexibly adapt the army to a particular enemy are the State's treasure.

Everything comes-goes, waxes-wanes and ebbs-flows. The only constant is change.

THE COMMENTARY

Our effective and successful whole ego-Self arrives first, rested and fresh and can take the initiative in an interpersonal conflict with an opposing whole ego-Self. 1) Some gainful advantage can be strategically extended, e.g., inviting an opponent to be comfortably seated in a more elevated position to give them a false sense of superiority and inviting them to have a beverage and 2) Some minor damage can be inflicted, e.g., spilling the beverage on them when serving it.

A strong, calm, comfortable and settled opposing whole ego-Self can be respectively acknowledged, stirred up, disquieted and moved about, e.g., through apologies and compliments and requests to change the lighting, temperature or seating arrangements.

Unexpected comments and moves can be made that throw an opposing whole ego-Self off balance and elicit responses from undefended places, e.g., making irrelevant comments about the meeting place, suddenly exiting for a bathroom.

Defensives of our whole ego-Self are insured by exhibiting behavior that cannot be attacked, e.g., at times agreeing with an opposing whole ego-Self. Offensives can be made by attacking undefended places of an opposing whole ego-Self, e.g., inquiring into an appearance of distress.

Our effective and successful whole ego-Self conducts itself in ways that make it impossible for an opposing whole ego-Self to know, when, where or how to defend or to attack, e.g., catching it off-guard and baffling it with unexpected, inconsistent, contradictory and unpredictable behavior.

Our effective and successful whole ego-Self remains a formless, mysterious and unknown entity so that an opposing whole ego-Self is incapable of understanding any of its plans, strategies and tactics or discerning its strong and weak points, e.g., by not being self-disclosing. This allows our whole ego-Self to have greater freedom and flexibility to act. The weak points of an opposing whole ego-Self are attacked and the strong points are avoided, e.g., by quickly and continually changing the focus of arguments.

The energy of our whole ego-Self is calmly conserved while that of an opposing whole ego-Self is wasted on all sorts of going this way and that in order to keep up and it ends up being imbalanced, thrown for a loop and defeating itself, e.g., as in martial arts like Ju-Jitsu, Judo and Aikido where the energy of the forward advancing movements of an opposing whole ego-Self are used to overthrow and defeat it.

If and when it is necessary or desirable to actively and directly engage an opposing whole ego-Self in interpersonal conflict, a number of strategies can be employed by our whole ego-Self. It can be forced out of defended positions into combat by attacking places that it must move to defend, e.g., criticizing certain points of their arguments.

It can be diverted from dealing with weighty and strongly opposed and conflicting issues, e.g., by first focusing on easier ones. It can be attacked on multiple fronts and will have to divide, disperse and weaken its forces, e.g., by bringing up several issues at one time. It can be weakened by capitalizing on its emotions and behavior, e.g., anger, greed and impulsiveness.

Such weakness and reduction in strength can result in an opposing whole ego-Self becoming reactive, discouraged and feeling that continuing to fight is futile.

If not being forced to fight or deciding not to fight, the arguments of an opposing and conflicting whole ego-Self can be thwarted by throwing it off course in some way, e.g., by asking for or taking a short break in the middle of the disputing.

The intensity of intrapersonal conflicts can be reduced by initially focusing upon some areas of agreement, minimizing differences and relenting somewhat as a foundation for gradually and progressively moving to address and deal with more difficult ones, e.g., 'seeing eye to eye' and 'giving in' on some issues.

The conflicting relationship can be conducted in a position and at a distance between opponents that is conducive, e.g., where and how far apart to comfortably sit.

An opposing and conflicting whole ego-Self can be prevented from successful fighting by discovering its intentions and plans, learning about its activities and movements, discerning its strengths and vulnerabilities and utilizing its own strategies and tactics, e,g., by intuiting motives and observing body language. The formlessness of our whole ego-Self is of critical importance for not allowing an opposing and conflicting whole ego-self to make similar discoveries and discernments, e.g., by not being self-revealing.

Direct strategic tactics leading to successful conflict resolution can be known and seen but it is the indirect ones that insure it, e.g., the former are used to deal with head-on confrontations and the strength of the latter lies in their concealment and surprise. Even when effective, both tactics are not fixed or used in every type of interpersonal conflict and are flexibly adjusted and varied according to its unique nature, circumstances and requirements.

Strategic tactics used in effectively dealing with and successfully resolving interpersonal conflicts are like the natural flowing of water from high to low places and taking its form from the ground upon which it flows. This means avoiding strong and engaging weak forces of an opposing and conflicting whole ego-Self and adjusting strategic tactics according to its formations.

Like water; the conditions, positions, formations, conduct, dispositions, movements, process and strugglings of interpersonal conflicts and their resolution do not have a constant or fixed shape. Being able to flexibly and appropriately modify and change strategic tactics to fit a particular opposing and conflicting whole ego-Self and specific contending issues is the genius that results in effectively dealing with and successfully resolving conflicting interpersonal relationships. Everything changes and invites and requires unique and flexible acting and responding.

The effectiveness and success of our whole ego-Self in proactively dealing with and resolving interpersonal conflicts involve:

1) preventative strategic tactics that quickly maneuver an opposing and conflicting whole ego-Self into weak, vulnerable and advantageous positions that can be quickly defeated or easily surrendered,

2) concealed, swift, surprise and decisive advances that catch an opposing and conflicting whole ego-Self off guard and quickly confuse, disrupt, disorganize and discombobulate it.

3) remaining formless and enigmatic so that plans, strategies and tactics cannot be known by an opposing and conflicting whole ego-Self and allow for the free, flexible and rapid implementation of strategic tactics and maneuvers,

4) relating to an opposing and conflicting whole ego-Self with dignity, humanity, respect and compassion,

5) being open to quickly resolving differences, disagreements, disputes, arguments and conflicts cooperatively and peacefully through diplomacy, negotiation, compromise or arbitration and

6) doing everything possible to end conflict quickly, early and without violating the viability, integrity and vitality of our whole ego-Self and without disrupting our human, moral and social order; without destroying our human lives, land and property and without desecrating and dispiriting our human Soul and Spirit.

> Using strategic invisibility
> and tactical flexibility
> maintain combat civility.

Your Reflections

CHAPTER SEVEN
MILITARY FIGHTING

軍　　　　爭

CHUN/JUN
Army/*Military*

CHENG/ZHENG
***Fight*/Contend**

CH'E/CHE – a cart/carriage
with axle, two wheels, body/
to roll/crush

+

MIEN/MIAN – roof/shelter

CHAO/ZHAO – claws/talons
to clutch/scratch/
'tooth and nail'

THE TEXT

SUN TZU SAYS:

In war, the Commanding General receives orders from the Ruler. An army is collected and forces are concentrated, harmoniously organized and encamped. Following that, comes the demanding and difficult task of strategic tactical maneuvering in the battle itself, the art of which is turning the devious into the direct and the disadvantageous into the advantageous. The difficulties involved are making distances circuitous, turning problems into advantages and reaching goals before the enemy, even when starting later.

Battling enemies has both advantages and dangers. Setting up a fully equipped army to seize an advantage will take too long but detaching a swift column involves abandoning equipment and provisions. Ordering soldiers to make long, forced and continuous marches in order to seize an advantage will result in the military leaders being captured. Stronger soldiers will arrive first, weaker ones will fall behind and only a few soldiers will reach the destination. Marching an army long distances to out-maneuver the enemy will lose division leaders and much of the army. The army will be lose its equipment, provisions and supply bases and will itself be lost.

Alliances cannot be formed without first knowing potential ally's designs. An army cannot be led without knowing the topography of the country, the lay of the land, its mountains, forests, rivers, marshes and swamps. Natural surroundings cannot be turned into advantages without using local guides and making appropriate decisions about moving troops.

Deception, calculated advantages and consolidating and ranking troops are required in order to succeed in warfare. Military forces must be as grand as forests, as swift as winds, as immovable as mountains and as raging as fire. Plans must be impenetrable and as dark as night and, when executed, as swift as lightning. Captured land and plundered spoils need to be divided among and allotted to soldiers of one's army and the people of one's country.

All army moves must be carefully evaluated, deliberated and planned. The conquering army will be practiced in both the artifice of indirect deception and the art of direct maneuvering.

On the battlefield, spoken words and ordinary signals are inadequate. Gongs, drums, flags and banners are necessary for focusing, coordinating, unifying and inspiring the army. In a single united army, the brave will not advance alone and the fearful will not retreat alone. To unite the army in night fighting, torches and drums are used and in day fighting, flags and banners are used. Doing so, the enemy army's Commanding General and the whole army will lose their presence of mind, heart, spirit and energy.

There are several arts of an effective General:

1) The art of studying moods and avoiding attacking an enemy army in the morning when the minds, spirits and energy of enemy soldiers are sharp and brisk and attacking later in the day when soldiers are more dulled and eager to return to camp.

2) The art of maintaining self-possession and patiently waiting for the appearance of any disorder and turmoil among the enemy soldiers.

3) The art of nurturing strength by being at ease while the enemy is struggling and being a well-fed army while the enemy is starving.

4) The art of studying circumstances by not engaging an enemy whose banners are in perfect order and whose formation is confidently gathered.

Advances against enemy armies are never made uphill and are never opposed when they are coming downhill. Enemy armies feigning retreat are not pursued. Enemy soldiers who are sharp are not attacked. Bait offered by the enemy is not taken. An enemy army returning home is not interfered with. A surrounded enemy army is given an outlet. A desperate enemy is not pressed.

Such are the strategic tactical arts of warfare and the skillful mastery of commanding, maneuvering and deploying army troops.

THE SYNOPSIS

In war, the General receives orders from the Ruler, collects, organizes and encamps the army.

Next comes the difficult task of tactical maneuvering, turning the devious into the direct.

It takes too long to fully outfit an army to fight but a swift one will sacrifice equipment.

Ordering soldiers to go on long marches will result in military leaders being captured.

Alliances must be made, the lay of the land assessed and local guides must be used.

Success in warfare depends upon deception, calculated advantages and unified troops.

Plans must be impenetrable, as dark as night and when executed, as swift as lightning.

Any and all army moves must be carefully evaluated, deliberated, planned and executed.

A winning army uses the artifice of indirect deception and the art of direct maneuvering.

In a united army, the brave will not advance alone and the fearful will not retreat alone.

Gongs, drums, banners and flags are used to focus the army and maintain *esprit de corps*.

The art of effective Generals is to avoid morning attacks when the enemy is alert and sharp,

to maintain self-composure and wait until the enemy shows disorder among its soldiers,

to stay strong by being at ease when the enemy struggles and well fed when they starve

and to not engage the enemy when banners are in order and troops confidently gathered.

The enemy is not advanced against uphill and is not opposed when coming downhill.

Enemy armies feigning retreat are not pursued and whose soldiers are sharp, not attacked.

Bait offered by the enemy is not taken and an army returning home is not interfered with.

A surrounded enemy army is given an outlet and a desperate enemy army is not pressed.

Such are the strategic arts of warfare and mastery of commanding and deploying an army.

THE COMMENTARY

In the beginning of interpersonal conflict, our True Tao-Nature manifests directives to our executive ego-self of our whole ego-Self which then: 1) collects, concentrates, organizes and secures our subordinate ego-selves of our whole ego-Self and then 2) sets about the challenging task of crafting strategic tactics to be employed in the conflict and to convert any disadvantges into advantages, e.g., proactively opening to our True Tao-Nature, integrating our executive and subordinate ego-selves in our whole ego-Self and then assessing the advantages and disadvantages and weighing the safety and danger and the benefits and risks of committing to a given interpersonal conflict prior to engaging in it.

Organizing and mobilizing our whole ego-Self to seize an advantage over an opposing whole ego-Self takes time but swiftly deploying some of our subordinate ego-selves leaves them unsupported, e.g., making a decision about either missing an opportunity or not fully gathering forces. To engage in an interpersonal conflict in distant places results in our whole ego-Self being exhausted and risks losing both our executive ego and our subordinate ego-selves, e.g., the stress and debilitating effects of traveling a long distance to settle a conflict ending up in defeat.

If support from allies is sought to assist in interpersonal conflicts and conflict resolution, their motives need to be known, e.g., are friends personally invested in helping for their own reasons. Our executive ego-self cannot effectively lead our subordinate ego-selves of our whole ego-Self into an interpersonal conflict without knowing and taking advantage of its situation and circumstances and without the guidance of those familiar with them who can assist in decision making, e.g., discussing the issues to be dealt with among those who have experienced similar ones and 'have been there, done that' before.

For our executive ego-self to succeed in interpersonal

conflicts and to achieve resolution; calculating advantages, consolidating our subordinate ego-selves and using secrecy and deception are necessary, e.g., forethought, planning, organization and subterfuge. The presence and force of our whole ego-Self must be as grand as forests, as swift as winds, as immovable as mountains and as overwhelming as fire. Plans must be dark as night, impenetrable and, when executed, as swift as lightning, e.g., being on solid ground and in high spirits, energetic and fully empowered when engaging opponents.

All strategies and tactics must be carefully planned, deliberated and studied, e.g., spending some time reviewing and rehearsing possible scenarios in an interpersonal conflict prior to meeting with opponents.

Learning something about opponents, their strengths and weaknesses, dispositions and preferences, etc., e.g., researching them on social media search engines.

Our effective and successful executive ego-self conflict resolver is well versed and practiced in the advantages and disadvantages of both direct and indirect strategic tactics and of avoiding strong opposition and moving to weak points, e.g., the skills involved in direct confrontation and the art of deception using both yang-yin energies.

The collective energy of our subordinate ego-selves is activated, united, consolidated and mobilized, e.g., by consciously and vividly reviewing, visualizing and integrating the finest, strongest and most unique characteristics, qualities and images of various key subordinate egos. Our whole ego-Self is presenced and readied in a radiant splendor that will astound and overwhelm opponents and discourage them from wanting to engage in fighting, e.g., showing up for a conflicted situation with high energy, full empowerment and a radiant presence.

In interpersonal conflicts, our effective and successful executive ego-self: 1) avoids engaging an opposing whole ego-Self early in the morning when spirits and energies are high, 2) calmly and patiently awaits indications of any disorganization among the executive ego-self and subordinate ego-selves of an opposing whole ego-Self, 3) conserves the energy, nurtures the strength and nourishes the bodies of our subordinate ego-selves while an opposing executive ego and subordinate ego-selves are using up energy, depleting strength and short of nourishment and 4) carefully studies the condition of the opposing forces and, if well-ordered and confidently gathered, does not advance against them, e.g., meeting an opponent later in the day, arriving early and awaiting their arrival at a place of your choosing, before their dinner and when they are low in energy, poorly organized and may want to go home.

In interpersonal conflicts, any advances against and engagements with an opposing whole ego-Self are not made from a lower position. A strong opposing whole ego-Self is not advanced upon. Bait offered by an opposing whole ego-Self is refused. An opposing whole ego-Self retreating or returning home is not followed or interfered with. A surrounded whole opposing ego-Self is given a way out. A desperate opposing whole ego-Self is not pressed, e.g., meet with an opponent in a restaurant, arrive first, sit in a higher seat, don't make the first move, ignore their saying that they will pay the bill, ask why they seem upset, don't stop them from leaving, show them the way out and don't follow them.

Such are some of the skillfully employed strategic tactical arts of the effective and successful managing and resolving of our interpersonal conflicts.

Right from the start
strategies are an art
straight from the heart.

Your Reflections

CHAPTER EIGHT
NINE VARIATIONS

九　變

CHIU/JIU
Nine/Numerous

P'IEH/PIE – left
downstroke/action
+
I/YI – one/second

PIEN/BIAN
Change/*Transform*

P'U/PU – tap/rap/knock
+
LUAN/LUAN – adjust/wrangle

THE TEXT

SUN TZU SAYS:

In war, the Commanding General receives orders from the Ruler, collects the army and concentrates military forces. Encampment is not made on unsuitable land. Alliances can be made in country where high roads intersect. Dangerously isolated positions are avoided. In hemmed-in situations, strategies must be resorted to. In desperate situations, fighting is necessary.

In war, there are roadways that should not be taken, armies that should not be attacked, towns that should not be besieged, territory that should not be contested and even commands of the Ruler that should not be obeyed.

A General who understands the advantages of strategic tactical variation, how to take advantage of terrains and how to manage soldiers knows how to effectively deploy soldiers and to successfully wage war. Some Generals know land configurations well but without knowing what should and should not be done will be unable to put that knowledge into practice. And those Generals who know the five constant factors well but are unversed in transformations and varying plans will not make optimal use of soldiers.

In the wise General's plans, advantages and disadvantages are integrated. This enables the effective implementation and successful accomplishing of the essential parts of strategic tactical plans, resolving difficulties, defeating the enemy and not endangering the army.

The power of enemy military leaders can be reduced by creating disadvantages, inflicting minor damages, keeping them constantly engaged, offering deceptive attractions and making them rush to defend a variety of places.

The *Art of War* is not relying upon the likelihood of the enemy not coming but rather upon being ready to receive them and not on the chance of them not attacking, but rather upon having made positions unassailable.

Five dangerous weaknesses of a Commanding General that can result in defeat in war are:

1) Recklessness, that leads to death, 2) Cowardice, that leads to capture, 3) Hot temper, that is easily provoked, 4) Scrupulous honor, that is easily disgraced and 5) excessive kindness to soldiers, that leads to trouble.

These are five weaknesses that are ruinous to effective leadership and successful conduct in warfare. Whenever an army is defeated and its Commanding General is killed, the cause will inevitably be found as due to these five dangerous weaknesses.

Let them be carefully considered, examined, studied, contemplated and reflected and meditated upon.

THE SYNOPSIS

In war, Generals receive orders from the Ruler and collect and concentrate the army forces.

Camps are not made in unsuitable country. Alliances are made where high roads intersect.

Isolated positions are avoided. In hemmed-in situations, strategies must be resorted to.

In desperate situations, fighting is necessary. Some orders of the Ruler are not obeyed.

Some roadways are not used, some armies are not attacked and some towns are not besieged.

Effective Generals know the advantages of terrain, tactical variations and use of troops.

Effective Generals integrate advantages and disadvantages and thus plan and implement successful strategies, defeat the enemy's army and not endanger or lose their own.

The power of an enemy army can be reduced by giving them trouble, inflicting damage, keeping them engaged, offering deceptive baits and making them rush to defend places.

The *Art of War* is not relying upon the enemy not coming but upon being ready for them and not relying upon chances of them not attacking but on making positions unassailable.

Faults of Generals can be recklessness leading to death, cowardice leading to capture, hot temper easily provoked, scrupulous honor easily disgraced, being overly kind to soldiers.

When an army is defeated or its General killed, it can be found to be due to these faults.

Therefore, to be undefeated, they must be carefully considered, examined and studied.

THE COMMENTARY

In interpersonal conflict, our executive ego-self has received a directive from our True Tao-Nature, has gathered up, concentrated and organized our subordinate ego-selves of our whole ego-Self and has positioned them in suitable conditions and circumstances that are not isolated, e.g., the leader, our executive ego-self, is accorded with our True Tao-Nature and integrates our various subordinate ego-selves of our whole ego-Self.

If and when the conditions and circumstances of an interpersonal conflict are restricting, strategies and tactics are planned, e.g., appealing to rational thinking, working with expressing feelings or changing behavior. If and when conditions and circumstances are desperate, fighting is required, e.g., using overt and direct confrontation.

In dealing with and resolving interpersonal conflict, there are: 1) paths that should not be taken, 2) areas that should not be besieged, 3) territory that should not be contested and 4) even directives from our True Tao-Nature that should be ignored, e.g., making false reports to police, pressing an opponent's ego-defenses with irrational demands, attacking character traits and questioning guidance from the universe or God.

Our effective and successful executive ego-self: 1) knows the atmosphere and terrain of interpersonal conflict and how to use the qualities and skills of our various subordinate ego-selves and 2) utilizes varied strategic tactics that are flexibly and specifically adapted to particular and uniquely shifting interpersonal conflicts, e.g., having the caring and angry subpersonalities of spouses dialogue with each other in marital conflict.

Just knowing the atmosphere and terrain of interpersonal conflict cannot be effectively and successfully put it into practice without also knowing what strategic tactics are appropriate to employ, e.g., not using intense cathartic techniques for domestic conflicts between an elderly couple.

Our executive ego-self knowing the five constants of Tao, heaven, earth, its own best qualities and disciplined organization but if being unversed in applying strategic tactical variations will not be able to make optimal use of our subordinate ego-selves in dealing with and resolving interpersonal conflicts, e.g., intellectually understanding Taoist philosophy and human psychology but being unpracticed in appropriately using meditation and psychotherapy methods, including dialoguing and role-playing.

Our effective and successful executive ego-self considers advantages and disadvantages in dealing with and resolving interpersonal conflicts, e.g., if not advantageous and possibly disadvantageous; not focusing upon certain problematic issues, not necessarily focusing on transference projections and power struggles and not using certain inappropriate psychotherapy intervention strategies and techniques.

Doing so is instrumental in effectively dealing with and successfully resolving conflicts without endangering our whole ego-Self and our subordinate ego-selves, e.g. successfully employing relevant and appropriate intervention strategies.

The power of an opposing and conflicting whole ego-Self of human beings can be defused, deflected, diffused and dispersed by keeping it engaged, encouraging it, deceptively enticing it and causing random activity, e.g., giving distracting and diverting messages to an opposing and conflicting whole ego-Self.

In the Tao/Way/Art of interpersonal conflict, our effective and successful executive ego-self does not rely upon: 1) the likelihood of an opposing whole ego-Self not appearing, but rather upon being prepared to meet it and 2) not on the chance that an opposing whole ego-Self will attack, but rather upon having made positions unassailable, e.g., being well grounded, centered, defended and prepared when awaiting or engaging in conflicted interpersonal relationships.

Five weaknesses of an ineffective and unsuccessful executive ego-self that are dangerous and ruinous to the conduct, positive outcome and resolution of interpersonal conflict with an opposing whole ego-Self are:

1) Recklessness leading to the negation of our executive ego-self and defeat of our whole ego-Self, e.g., being careless.

2) Cowardice leading to capture and being defeated by and opposing whole ego-Self, e.g., being timid.

3) Hot temper leading to easy provocation by an opposing ego-Self and ultimate defeat, e.g., being ardent.

4) Scrupulous honor leading to easy disgrace by an opposing ego-Self and ultimate defeat, e.g., being ingratiating.

5) Excessive attachment to and kindness toward our subordinate ego-selves leading to defeat, e.g., being over-identified.

Whenever interpersonal conflict is not effectively dealt with and successfully resolved and our executive ego-self is negated and our whole ego-Self is defeated, it can be attributed to these five weaknesses of our executive ego-self, e.g., if psychotherapists or counselors are careless, insecure, unstable, fastidious and over-identified with patients or clients and need their own therapy and/or peer consultation and supervision in order to be competent practitioners.

Because these issues are critical to effectively dealing with and successfully resolving interpersonal conflict, they need to be carefully considered, examined, studied, contemplated and reflected and meditated upon by our executive ego-self of our whole ego-Self, e.g., ongoing thorough self-reflection, self-examination, self-understanding and self-correction are critical for being an ethical, responsible, accountable and competent leader.

CLOSING COMMENT

This concludes Chapters Five, Six, Seven and Eight on interpersonal relationships and conflict. Interpersonal conflict is successfully resolved by, through, in and as the harmonious co-existing and cooperative interrelationship of our whole ego-Self with that of a formerly opposing whole ego-Self. As such, this preserves the original unity, integrity, peacefulness and pacifism of our True Tao-Nature and Tao-State within our awakened, enlightened and liberated consciousness, conscious awareness and conscious experience of our interpersonal relationships.

The general's fault
results in enemy assault
and brings victory to a halt.

Borders . . . . . .
the vibrant dynamic interfaces
and semi-permeable membranes
between international wars
and peaceful graces.

Boundaries . . . . . .
the vibrant dynamic spaces
and connecting distances
between interpersonal conflicts
and resolving embraces.

SOME EXAMPLES

Since the commentaries are also generalized abstractions of interpersonal relationships, it may be helpful to provide concrete examples of the relationships between our executive ego-self/I, subordinate ego-selves/me-s and whole ego-Self/human being and those of opposing and conflicting ones.

1) Our executive ego-self/I may identify with and deploy its courageous sub-personality/me to defend against an aggressive sub-personality of an opposing whole ego-Self in order to successfully protect the integrity of our whole ego-Self/human being.

2) Our executive ego-self/I may identify with and engage its wise sub-personality/me with the executive ego-self of an opposing whole ego-Self to negotiate a conflict resolution and successfully sustain the integrity of our whole ego-Self/human being.

YOUR REFLECTIONS

NON-PERSONAL
RELATIONSHIPS

LANDSCAPE

CHAPTER NINE

ARMY CONDUCT

行　　軍

HSING/XING
Go/March

CHUN/JUN
Army/Military

CH'IH/CHI – walk/
step with right foot
+
CH'U/CHU – bring
left foot forward

CH'E/CHE – a cart/carriage/
to roll/crush
+
MIEN/MIAN – roof/
shelter

THE TEXT

SUN TZU SAYS:

The following are directives regarding the encampment and positioning of the army and for observing and confronting enemy armies:

1) For mountain warfare: mountains are quickly crossed and valleys are used. Camping should be in high places and facing the sun. Heights are not to be climbed in order to fight.

2) For river warfare: the army is not placed close to rivers. After crossing a river, it should be left quickly. Do not advance to meet an enemy army in mid-stream. Let half of the enemy army cross and then attack. Do not go to meet an enemy army near a river's edge that it has to cross. Do not move upstream to meet an enemy army. Due to heavy rains, if a river needing to be forded is flooded, wait until it recedes.

3) For marsh and swamp warfare: cross them quickly. If

forced to fight in them, stay near fresh water and in grasses with your back to trees.

4) For flat country: take a position with rising ground behind so that danger is in front and safety is behind.

These are four useful kinds of military knowledge that have been found to be effective for successfully winning wars and defeating enemy armies.

Armies generally prefer high ground and sunny places (yang) to those that are low and dark (yin). Camping on solid ground is vitalizing and healthy. When setting up defenses, the sunny side of a hill is occupied with the slope at the rear. Doing so, both supports soldiers and uses the natural advantages of the terrain.

Country in which there are precipitous cliffs with channels running between them; deep hollows, fissures and crevices; confined places and tangled thickets should not be approached or should be left quickly. A strategic advantage is obtained by getting the enemy army to enter them.

In hilly country, near ponds surrounded by grass, in hollow basins filled with reeds or in woods with dense undergrowth, they all should be searched out for enemy soldiers lying in ambush or spies in hiding.

When the enemy is close by and quiet, they are depending upon the strength of their position. When the enemy is eager to be being advanced upon, they try to provoke a battle. When the enemy encampment is easily accessible, they are baiting.

Movement of forest trees indicates that the enemy army is advancing. Obstructions in thick grass means that the enemy wants to arouse doubts. Birds flying up is a sign of an ambush. Startled animals indicate that a sudden attack is coming.

Dust rising in high columns is a sign that chariots are advancing. When the dust is low and spread over a wide area it means that infantry is approaching. Thin shafts of scattered dust indicate the gathering of firewood. A few clouds of dust coming and going indicate that the army is encamping.

The deferential speaking of enemy envoys is a sign of increased preparation for an advance and impending attack. Aggressive speaking and lunging forward as if to attack means that they are going to retreat. When the light chariots come out first and take a position on the wings, it indicates that the enemy is forming for battle and readying to attack. Peace proposals without advanced notice and prior arrangements suggest a plot.

When enemy soldiers are running around and quickly falling into rank, it means that the critical moment of engaging in warfare has arrived. When some enemy soldiers are seen advancing and some are seen retreating, it is a lure. When enemy soldiers are seen standing and leaning on their spears, it is a sign of being faint from hunger. If those enemy soldiers sent for water begin drinking it first by themselves, their army is suffering from thirst.

If an enemy army fails to take an obvious advantage, it means that their soldiers are exhausted. When birds gather at any place, it means that it is unoccupied. If noisy shouting is heard, it indicates anxiety and fear. Disturbances in camp indicate that a Commanding General's authority is weak. Banners and flags randomly moving around means chaos. Angry officers indicate that they and the soldiers are fatigued.

When an enemy army feeds its horses with its own grain, slaughters its own cattle for food, fails to hang up its cookware and does not return to its tents; it means that they are desperate and will be fighting to the death. Soldiers gathered together in small groups and speaking in low tones indicates dissension within the ranks.

Frequent rewards and punishments show that enemy Generals are in extreme crisis and distress. If they begin with bravado and then fear the enemy army, they lack intelligence and competence. When enemy envoys come giving compliments and seeking arrangements, it is a sign that a truce is desired.

If enemy troops angrily advance and face off for a lengthy

time without starting a battle, their motives require cautious vigilance and careful examination. If enemy troops are larger in number, their available strength is concentrated and soldiers are inspired; the enemy army needs to be closely watched and reinforcements need to be sent for.

A General who has no forethought and underestimates the enemy will be captured by them. When soldiers are punished before establishing an attachment to superior military officers, they will not be subordinate and will be difficult to use in battle. If they do establish an attachment but punishments are not enforced, they also will be relatively useless in battle.

Military soldiers must be treated humanely but also kept under control by strict discipline. When commands are regularly and consistently enforced, the army will be well-disciplined and strong and the soldiers will be useful. When a Commanding General genuinely exhibits confidence in soldiers and orders are obeyed, mutual trust and benefit ensue.

THE SYNOPSIS

For encamping and positioning an army and for observing and engaging the enemy army:

Mountains are avoided, valleys are taken and camping is made in high places facing the sun.

River crossing is quick and the enemy is not met in rivers hard to cross or met in mid-stream.

Let half of the enemy army cross a river and then attack. Do not advance upstream to meet it.

Cross marshes quickly. If forced to fight there, have water and grass nearby and trees behind.

On level land, have rising ground to the right and rear so danger is in front and safety behind.

Armies prefer high ground, sunlit places, occupying a sunny hillside sloping on the right rear.

Precipitous cliffs, narrow crevices, confined places and dense thickets should be avoided.

Strategic tactical advantage is gained by forcing an enemy army to approach and enter them.

Hollow basins and ponds with tall grass and dense forests are searched out for ambushes.

A close quiet enemy is strong, a provocative enemy is ready, an accessible enemy is baiting.

Moving trees indicate an advancing army. Obstructions in tall grasses are to arouse doubts.

Flying birds indicate an ambush. Startled animals indicate a sudden enemy attack is coming.

High dust columns indicate the movement of chariots. Low ones, indicate troop movement.

Thin shafts of scattered dust indicate firewood collecting. Transient clouds indicate camping.

Deferential or aggressive enemy talking respectively indicates either an advance or retreat.

Chariots assuming wing positions indicate a coming attack and peace proposals are a ploy.

When enemy soldiers are milling around and falling into rank, it means immanent attack.

Advancing and retreating soldiers is a lure and soldiers leaning on spears are starving.

When soldiers sent for water are first drinking it themselves, it indicates a general thirst.

An army failing to take an advantage is exhausted. Shouting indicates anxiety and fear.

Camp disturbances indicate weak command. Angry officers indicate they are fatigued.

Banners and flags being shifted around indicate that a troop insurrection is immanent.

Feeding horses one's grain, slaughtering cows for food, not hanging up pots and not going back to tents indicate that enemy soldiers are desperate and will likely fight to the death.

Soldiers in small groups whispering low to each other indicates dissension among them.

Frequent rewards and punishments indicate that a General is in crisis and distressed.

Beginning with bravado and then being frightened of the enemy army is incompetence.

When enemy envoys approach with compliments, it indicates that a truce is desired.

When enemy troops angrily march up and then stand still, vigilance is then needed.

When one's troops are fewer in number, no attack is made and reinforcements sent for.

A General without foresight and who underestimates the enemy will be captured.

Solidiers punished before attachment to superiors will be insubordinate and useless.

If soldiers do form attachment but punishments are not enforced, they will be useless.

Soldiers must first be treated humanely but also must be controlled by strict discipline.

When commands are enforced, the army will be well disciplined and soldiers useful.

When Generals show genuine confidence in soldiers and orders are obeyed, mutual trust and benefit ensue.

THE COMMENTARY

This chapter and Chapters 10, 11 and 12 focus upon the non-personal atmospheric conditions, topographic grounds, surroundings and situations, settings and features of varying advantageous and disadvantageous environments of conflict and the general characteristics, configurations and behavior of an opposing whole ego-Self.

Our executive ego-self takes positions that avoid pitfalls and drawbacks and from which an opposing whole ego-Self can be observed, evaluated and engaged or not, e.g., it is important to take safe and advantageous positions to observe opponents.

1) Insurmountable conflicts are quickly passed over and lower-order ones are focused upon, e.g., less complicated and involved issues are dealt with first.

2) Continuously fluid conflicts are met by not catching an opposing whole ego-Self mid-stream, e.g., letting the flowing of conflict take its course and move to its natural completion.

3) Bogged-down conflicts with an opposing whole ego-Self are quickly passed over, e.g., stalemated conflicts that are going nowhere are disregarded.

4) Evenly situated conflicts with an opposing whole ego-Self are confronted head-on, e.g., conflicts between equally strong whole ego-Selves are fully engaged in.

Our effective and successful executive ego-self and our subordinate ego-selves of our whole ego-Self adopt advantageous positions that are elevated, sunny and brightly illuminated yang ones that are above the lower, shady and dark yin ones, e.g., conflicts are best dealt with from the vantage point of the enlightened consciousness of our Higher Self and our True Tao-Nature.

Steep conflicts with schisms, abysmal conflicts, bogged-down conflicts, constricting conflicts and entangled conflicts are not engaged in by our effective and successful whole ego-Self but which can gain strategic advantages by enticing an opposing whole ego-Self to get lost in them, e.g., difficult, endless, stalemated, pointless and complicated issues can provide opportunities to divert and occupy an opposing whole ego-Self and deplete its energy and strength.

In relation to the ups and downs and ins and outs of conflicts with depressions and pitfalls to fall into or constricting complications to get caught up in, our effective and successful executive ego-self searches for the hidden motives of an opposing executive ego-self and subordinate ego-selves waiting in ambush, e.g., pauses in dealing with conflicts can provide opportunities to intuit the intentions of an opposing whole ego-Self.

In conflicts, if an opposing whole ego-Self is close and quiet, it probably is feeling strong; if it is provocative, it likely is eager to be confronted and challenged and if seems open and accessible, it probably is baiting, e.g., knowing the positions and actions of an opposing whole ego-Self enable understanding its strength and intentions.

Unified movements of an opposing whole ego-Self indicate a readiness to advance; blocking engagements are used to arouse doubts and suspicions; sudden flurries of energy signal that a head-on attack is coming and various smoke-screens indicate a solidifying power, attempting to hide and obscure activities and advancing to directly attack, e.g., observing the activities of an opposing whole ego-Self enables knowing its intentions and can assist in preparing for its impending advancing and immanent attacking.

Deferential speaking on the part of an opposing whole ego-Self likely indicates an impending advance and aggressive moving as if to attack likely precedes a retreat. When an opposing whole ego-Self brings up light side-issues, it is a likely sign of forming for battle and preparing to attack. If agreeing and disagreeing keep alternating, it likely is a lure. If attempting to make peace, it likely is plotting an attack. If the subordinate ego-selves of an opposing whole ego-Self are milling around and organizing themselves, the moment of direct full-on fighting has arrived, e.g., the behavior and movements of an opposing whole ego-Self and its subordinate ego-selves can reveal positioning for advances and retreats and attacks.

The body language of subordinate ego-selves and requests of an opposing whole ego-Self can indicate the state of their being, strength and energy, e.g., if they are fatigued, thirsty, hungry or exhausted. Missing out on obvious advantageous opportunities to gain the upper hand, indicates fatigue. Noisy shouting and disturbances indicate anxiety. Fear indicates a weak executive ego-self.

When order and calmness are out of control, disorganized and angered; the executive ego-self of an opposing whole ego-Self is exhausted and failure is immanent.

If the executive and subordinate ego-selves of an opposing whole ego-Self are neglecting their appearance, abandoning their usual decorum and failing to return to and maintain rationality; they are desperate and will be fighting to the end, e.g., an opposing whole ego-Self facing and readying for an immediate immanent battle, 'pulls out all of the stops' and neglects and abandons its customary conduct, activities and routines.

In the organizing and managing of an opposing whole ego-Self by its executive ego-self; if its subordinate ego-selves are fighting among themselves and are being given frequent rewards and punishments, it indicates a crisis and extreme distress. If an executive ego-self of a whole ego-Self begins conflict with bravado and then becomes fearful of an opposing whole ego-Self, it indicates an extreme lack of intelligence and competence, e.g., the strength and courage of a leader should be maintained throughout conflict.

If the envoys of an opposing whole ego-Self approach our executive ego-self with compliments, it is a sign that some kind of a truce is desired, e.g., peace-seeking gestures that involve expressions of respect and courtesy.

If an opposing executive ego-self and whole ego-Self angrily approach and face off without either advancing further or retreating, vigilance is required, e.g., in order to discern their next steps. If an opposing whole ego-Self has a fewer number of subordinate ego-selves, it means that no direct attack will be made. But if the opposing whole ego-Self has a larger number of subordinate ego-selves, the strength of our whole ego-Self needs to be consolidated, the opposing whole ego-Self needs to be closely observed and reinforcements need to be sent for, e,g, if it seems like defeat may potentially occur in a conflict, ego-strengths need to be consolidated and supportive friends may need to be called in to assist.

If our executive ego-self of our whole ego-Self does not have forethought and underestimates the strengths of an opposing whole ego-Self, it will likely be captured, e.g., not thinking ahead before a conflict and minimizing the strength and abilities of an opponent may result in being held in their control.

If our executive ego-self of our whole ego-Self attempts to organize our subordinate ego-selves by strict discipline before establishing a valued and inclusive relationship with them, they will be insubordinate and relatively useless for the harmony, strength and effectiveness of our whole ego-Self.

If a valued and inclusive relationship is made but discipline is not clearly and consistently enforced, our subordinate ego-selves will negatively affect the integrity and efficacy of our whole ego-Self and will also be relatively useless, e.g., the usefulness of our subordinate ego-selves in dealing with and resolving conflicts depends upon their valued inclusion and disciplined organization in our whole ego-Self.

Our executive ego-self of our whole ego-Self needs: 1) to genuinely, humanely and harmoniously integrate and consolidate and 2) to clearly, confidently, decisively and fully control and manage our various subordinate ego-selves in order for them to co-create the enabling power of, and to usefully participate in, our whole ego-Self's effectively dealing with and successfully resolving conflicts and experiencing the mutual benefits of doing so, e.g., our subordinate ego-selves need to be genuinely valued, harmoniously integrated and decisively disciplined.

Our effective and successful executive ego-self: 1) calmly observes and makes the most of various environmental conditions, situations and circumstances; 2) safely adopts positions and makes appropriate dispositions and 3) exercises humaneness, discipline, forethought, caution and prudence that result in the unity, harmony and utility of our subordinate ego-selves and that place an opposing whole ego-Self in positions of being rendered unthreatening, realizing the futility of conflicting, surrendering and thus being defeated without fighting.

For example: 1) our effective and successful leader of our whole ego-Self closely observes the situation of a conflict and capitalizes upon its advantages; 2) safely strategically positions and tactically employs our harmoniously unified subpersonalities of our whole ego-Self and 3) utilizes them in ways that render an opposing and conflicting whole ego-Self benign and to thus be defeated without fighting.

> Victories are served
> and human lives preserved
> by enemy conduct observed.

Your Reflections

CHAPTER TEN
TERRAIN FORMS

地　　形

TI/DI
Ground/Place

T'U/TU – earth/place
+
YEH/YE – utensil/also/still

HSING/XING
Form/Figure

SHAN/SHAN – feathers
+
CH'IEN/QIAN – two shields

THE TEXT

SUN TZU SAYS:

The six kinds of terrain or ground and situational positions are: 1) Accessible, 2) Entangling, 3) Stalemated, 4) Narrow, 5) Precipitous and 6) Distant.

1) Accessible ground can be freely traveled and occupied by both armies. When first to occupy, raised and sunny spots are found and supply lines are carefully guarded for the advantage in warfare.

2) Entangling ground can be entered but is difficult to return to. From this position, when the enemy is unprepared, they can be defeated but when they are prepared and are not defeated, a return is impossible.

3) Stalemated ground is when the position is one where neither side gains by making the first move. Baiting by the enemy

army should be ignored and retreating is advisable to entice the enemy to come forward and to then attack when part of their army advances.

4) Narrow ground or passes can be occupied when first to do so and need to be strongly held while awaiting the enemy army. If the enemy army arrives first and prevents occupation of a pass, they are not pursued unless the pass is not strongly held.

5) Precipitous ground can also be occupied when first to do so and raised and sunny spots should be chosen to lie in wait for the enemy army. If the enemy army has occupied first, they are not engaged and a retreat is made in an attempt to lure them away.

6) Distant ground is being situated far away from the enemy army when the strength of both armies is equal and provoking and fighting a battle is disadvantageous.

All six of these principles are connected to earth and using terrain and are the utmost responsibility of the Commanding General to carefully study them.

An army can suffer the following six calamities aris-ing not from natural causes but from errors attributable to the Commanding General: 1) Flight, 2) Insubordination, 3) Collapse, 4) Ruin, 5) Chaotic disorder and 6) Disastrous defeat.

1) Flight happens when one military force is pitted against another that is ten times its size.

2) Insubordination is when the soldiers are too strong and the military officers are too weak.

3) Collapse is when the military officers are too strong and the soldiers are too weak.

4) Ruin occurs when military officers are angry and attack on their own before the Commanding General can ascertain whether or not the army is in a position to fight.

5) Chaotic disorder is when a Commanding General is weak and without authority, orders are not clear, no fixed duties are assigned to military officers and soldiers and ranks are formed in a haphazard manner.

6) Disastrous defeat occurs when a Commanding General permits a smaller or weaker military force to engage a larger or stronger one and does not place elite soldiers in the front rank.

As with the six terrains, these six calamities also are the utmost responsibility and duty of the Commanding General and also must be carefully studied.

The natural formations of a country are an army's critical support but the power of estimating the enemy, controlling the forces of conflict and calculating its difficulties and dangers are the characteristics of a successful General. When this knowledge is put into practice in fighting, battles will be won and if not they will be lost.

If fighting will surely end in victory, then it must continue even though a Ruler forbids it. If fighting will surely end in defeat, then it must not be fought even though a Ruler orders it.

A General who advances without desiring fame and who retreats without fearing shame and whose only interest is in safeguarding the country and serving the Ruler well is the treasure of the State.

When military soldiers are regarded as one's own beloved sons they will follow into the deepest valleys and stand by until the death. However, in spite of kindness, if authority is not also felt, commands are not enforced, control is not maintained and disorder is not stopped; the soldiers will behave like spoiled children and are relatively useless for warfare.

Knowing that military soldiers are ready for battle but not knowing that the enemy is not open to attack or knowing that the enemy is open to attack but that soldiers are not ready for battle is being victorious only half of the time. Knowing that soldiers are ready for, and the enemy is open to, attack but not knowing that the battleground terrain makes fighting impracticable also is being victorious only half of the time.

Those experienced in warfare, when moving, are never confused or erroneous and, when acting, are never at a loss or limited. When you know yourself, the enemy, heaven and earth; victory will be certain and complete.

THE SYNOPSIS

The six types of terrain, ground or situational positions are:

1) Accessible ground that can be freely traveled, occupied and utilized by both armies.

2) Entangling ground that can be fought on but can be lost and return to it made impossible.

3) Stalemated ground where neither army gains anything by making the first move.

4) Narrow ground that can be occupied by the first army to do so but must be safeguarded.

5) Precipitous ground that can be occupied by the first army and sunny high spots chosen.

6) Distant ground that is far away from the enemy and when both of the armies are strong.

All of these situations are a General's responsibility to know and to carefully study.

Six types of calamity can befall an army that are due to errors of a General:

1) Flight occurs when the enemy army is ten times more in size and strength.

2) Insubordination occurs when soldiers are too strong and army officers too weak.

3) Collapse occurs when army officers are too strong and soldiers are too weak.

4) Ruin occurs when angry officers fight on their own before a General orders it.

5) Chaos is when a General has no authority, orders are unclear, ranks are unfixed.

6) Disastrous defeat is when a General allows a weaker army to fight a stronger one.

All of these errors also are a General's responsibility to know and to carefully study.

Terrain formations are a great asset and the General's responsibility is a critical factor in estimating the enemy, controlling forces and calculating difficulties and dangers.

When this knowledge is put into practice, enemies will be defeated and wars won.

When wars can be won, they are fought even when not being ordered by the Ruler.

When wars cannot be won, they are not fought even when ordered by the Ruler.

A treasured general fights not for fame but to protect the State and serve the Ruler.

Soldiers regarded as sons will follow into dangerous terrain and fight to the death.

Soldiers so regarded will behave like spoiled children if not consistently disciplined.

Victories are achieved half the time when soldiers are battle-ready but the enemy is not open to attack, when the enemy army is open to attack but soldiers are not battle-ready and when both are ready and open but the battleground is impracticable for fighting.

Experienced warfarers are not confused or in error when moving and not at a loss or limited when acting.

Knowing oneself, the enemy and heaven and earth, victory will be certain and complete.

THE COMMENTARY

In the settings and fields of conflict; six kinds of terrain, grounds, situations and environments; their dangers, advantages and disadvantages and the positions, options and strategic tactical maneuvers of a whole ego-Self in relation to them are:

1) Accessible – a place of conflict that both our whole ego-Self

and an opposing whole ego-Self can occupy, e.g., a space that is neutral or a shared position in a conflict that is not advantageous for either whole ego-Self. Advantages are obtained by safeguarding strengths and quickly taking an elevated and illuminated position of our True Tao-Nature and Higher Self in the conflict.

2) Entangling – a place of conflict that can be exited from but that is difficult to re-occupy, e.g., abandoning a committed but ineffective position in a conflict that, once left, is difficult to return to and re-institute if needing to.

3) Stalemated – an initial impasse situation of conflict where neither whole ego-Self gains by making the first move, e.g., invitations from an opposing whole ego-Self to begin are declined. A retreat from engaging can induce the opposing whole ego-Self to come forward with a subordinate ego-self that can be easily dealt with.

4) Narrow – tight spots in a conflict that can be strongly held onto until an opposing whole ego-Self comes forward in some way, e.g., an opposing whole ego-Self prevents holding the tight spots, no pursuit or confrontation is made unless its position is not strongly held.

5) Precipitous – the ground needs to be occupied first and high and sunny spaces taken to await the moves of an opposing whole ego-Self, e.g., again, quickly assuming an elevated and illuminated position of our True Tao-Nature and Higher Self in the conflict affords an advantage. If the opposing whole ego-Self is first to occupy the space, forestall engaging, retreat and attempt to draw it in.

6) Distant – this ground of conflict is situated far away from an opposing whole ego-Self and when both whole ego-Selves are equally strong. Engaging in a conflict or fighting a battle is not advantageous and is not done, e.g., end the long-distance cell-phone argument by 'hanging up'.

All six of these grounds and situations and positions and strategic tactics are important responsibilities of our executive ego-self of our whole ego-Self in effectively dealing with and successfully resolving conflicts and they need to be fully understood.

Our whole ego-Self can suffer the following six calamities that are attributable to the leadership faults and strategic tactical errors of our executive ego-self:

1) Flight – occurs when our whole ego-Self is pitted against an opposing whole ego-Self ten times its strength, e.g., don't attempt to engage in conflict with someone obviously much stronger than yourself.

2) Insubordination – occurs when our subordinate ego-selves are very much stronger than our weaker executive ego-self, e.g., don't allow any of your subpersonalities to take charge of your executive ego-self.

3) Collapse – occurs when our executive ego-self is very much stronger than our weaker subordinate ego-selves, e.g., don't let your executive ego-self lord its superiority over your subpersonalities.

4) Ruin – occurs when our executive ego-self is angry and impulsively begins fighting without knowing whether our subordinate ego-selves are in a position to fight, e.g., don't lose touch with the accessibility, strengths, weaknesses and readiness of your subpersonalities.

5) Chaos – occurs when our executive ego-self is weak and without authority, gives unclear orders to, and assigns no fixed duties to, our subordinate ego-selves and forms their ranks in a haphazard manner, e.g., don't have unauthoritative, unclear and disorganized role-relationships with your subpersonalities.

6) Defeat – occurs when our executive ego-self allows its smaller and weaker whole ego-Self to engage a larger and stronger opposing whole ego-Self and without using our elite subordinate ego-selves on the frontlines of a conflict. e.g., don't engage in conflict with an obviously stronger person and especially not without having the strongest subpersonalities at the forefront of your conflicted encounter.

As with grounds and situations and positions and strategic tactics; all six of these faults and errors are important responsibilities of our executive ego-self of our whole ego-Self in effectively dealing with and successfully resolving conflicts and they need to be fully understood.

Knowing the situational landscape of a conflict and its configurations are a great advantage, but the ability to estimate the strength of an opposing whole ego-Self and to calculate dangers and difficulties are characteristics of our effective and successful executive ego-self.

When put into practice, conflicts will be won and resolved and, when not, conflicts will be lost and remain unresolved, e.g., a conflict situation may be advantageous but effectively dealing with and successfully resolving a conflict will not occur without knowing the dangers of the situation and the strengths of an opposing whole ego-Self.

When a conflict will surely and clearly be resolved and won, it should continue even if it is obscuring the presence of our True Tao-Nature in the field of conscious awareness of our executive ego-self of our whole ego-Self, e.g., only conflicts resulting in resolution are engaged in and continued even when seemingly contrary to the True Tao-Nature and Higher Self of our whole ego-Self.

When a conflict will surely and clearly not be resolved and end in defeat, it should not continue even when it may seem to serve sustaining the presence of our True Tao-Nature in the field of conscious awareness of our executive ego-self, e.g., conflicts resulting in defeat are not engaged in or continued even when seemingly in accord with the True Tao-Nature and Higher Self of our whole ego-Self.

Our executive ego-self that makes positive advances in conflict resolution only attends to the matter at hand without desiring fame and that retreats in conflicts without fearing shame; and whose only interest is in serving, safeguarding and preserving the Tao-State, our True Tao-Nature, our whole ego-Self and our subordinate ego-selves, is the treasure of the Tao-State, e.g., our leader is a treasure of the Tao-State, our True Tao-Nature and our whole ego-Self when free of personal egocentric motivations and when solely concerned with maintaining and sustaining that Tao-State and our True Tao-Nature.

When our executive ego-self regards and values our subordinate ego-selves as its own reflections and embodiments; they will be harmoniously integrated, fully compliant and dutifully follow into the most dangerous and life-threatening conflicts, e.g., this is the invulnerability of our fully integrated whole ego-Self.

However, in spite of such true and deep kindness; if authority is not exercised, discipline is not given, commands are not enforced and disorder is not stopped; our subordinate ego-selves will act out in various ways that render them useless for assisting in winning and resolving conflicts, e.g., in spite of the embodiment and harmonious integration of our subordinate ego-selves by our executive ego-self to safeguard and preserve our whole ego-Self and True Tao-Nature, they need to be well-disciplined.

Only half of conflicts are effectively dealt with and successfully resolved: 1) when our subordinate ego-selves of our whole ego-Self are ready to deal with conflicts but opposing whole ego-Selves are not open to doing so, 2) when opposing whole ego-Selves are ready for dealing with conflicts but our subordinate ego-selves of our whole ego-Self are not ready for doing so and 3) when the atmospheric conditions and topographic situations of the conflicting field of conscious awareness are not practical or suitable, e.g., effectively dealing with and successfully and fully resolving conflicts involves both our subordinate ego-selves and opposing ones being ready and open to do so and conditions and situations being suitable.

Our effective and successful executive ego-self of our whole ego-Self; when positioning and deploying our subordinate ego-selves and when making strategic tactical maneuvers; is flexibly unlimited in specific relevant possibilities and is not confused, at a loss or erroneous in actual engagements with opposing whole ego-Selves, e.g., in successful conflict resolution; clear, certain, precise and flexible strategic tactical maneuvering of subpersonalities in relation to changing circumstances is essential for success.

When our executive ego-self of our whole ego-Self: 1) remembers and fully identifies with and as original, absolute, ultimate, essential and inborn True Tao-Nature and Heaven-Earth and 2) thoroughly knows and accepts itself, 3) fully and harmoniously integrates our subordinate ego-selves and 4) clearly and fully understands the nature and movements of an opposing whole ego-Self; certain and complete victories will be won when dealing with and resolving conflicts.

> When the enemy is known,
> strategies are shown
> and victories are one's own.

YOUR REFLECTIONS

CHAPTER ELEVEN

NINE SITUATIONS

九　地

CHIU/JIU
Nine/Numerous

P'IEH/PIE – left
downstroke/action
+
I/YI – one/second

TI/DI
Ground/*Situation*

T'U/TU – earth/place
+
YEH/YE – utensil/also/still

THE TEXT

SUN TZU SAYS:

The *Art of War* recognizes nine grounds, terrains or situations which are:

1) Dispersing ground – when the army is fighting in its own territory. Do not fight.

2) Facile ground – when the army has entered into enemy territory. Do not linger.

3) Contentious ground – possession of which is mutually advantageous. Do not attack.

4) Open ground – both sides have freedom of movement. Do not become separated.

5) Intersecting ground – triply connected and first arrival commands. Join with allies.

6) Serious ground – when the army has penetrated deep into enemy country. Plunder.

7) Difficult ground – when terrain is difficult to traverse. March swiftly and steadily.

8) Hemmed-in ground – with narrow gorges and tortuous exits. Resort to strategy.

9) Desperate ground – where fighting is the only way to not be destroyed and die. Fight.

The skillful General knows how to separate the enemy's front and rear lines, to disrupt coordination and support between large and small divisions and to hinder officers and soldiers from communicating and cooperating with each other.

When enemy soldiers are united they are kept in disorder. When a forward move is advantageous, it is made otherwise remain still. When an orderly array of enemy soldiers is marching to attack, something is seized that is of value to them.

Speed is the essence of war. Take advantage of the enemy's unpreparedness, make way by unanticipated and unexpected routes and attack unguarded spots.

The following are principles to be observed by an invading military force:

1) Further penetration into enemy territory results in greater troop solidity.

2) Forays are made into fertile country in order to supply the army with food.

3) Carefully monitor the morale and well-being of soldiers and do not overtax them.

4) Concentrate and conserve energy and strength. Devise unpredictable strategies.

5) Put soldiers in positions where there is no escape so they will fight with utmost strength instead of flee. Soldiers in desperation lose a sense of fear and when there is no place of safe refuge or options, they will stand firm and fight hard to the end. Soldiers will be faithful, dutiful, dedicated and trusting without being ranked, positioned or given orders.

6) Prohibit the making of omens and dispel superstitious

doubts to avoid fears of calamity.

If military soldiers relinquish money, it is not because of eschewing wealth and if their lives are short it is not because of abhorring longevity.

On the day of battle, soldiers may cry profusely but once in battle they will display the courage of famous warriors of old and the fight of a legendary snake, Shui-Jan/spontaneously responsive, that counterattacks with head or tail when either are attacked or with both when the middle is attacked. An army can be like this snake.

Enemy soldiers caught in the same dire natural events will become supporting and assisting.

In warfare, it is insufficient to just tether horses and bury chariot wheels to maintain order. An army is unified by setting one uniform standard of courage that all soldiers reach, share and enact in unison. Leadership is exercised, the most is made of both strong and weak points and the correct use of different terrain is made.

Skillful Generals conduct the army as though leading a single soldier by the hand. They are calm and quiet to ensure secrecy and upright and fair to maintain order. They mystify officers and soldiers with false reports and appearances to keep them unknowing and spontaneous. By changing actions and revising plans, they keep the enemy without definite knowledge and by shifting campsites and taking circuitous routes, they prevent the enemy from knowing locations and anticipating moves.

At critical moments, Generals act like someone who has soldiers climb high up on a ladder and then kicked it away. They lead their soldiers deep into enemy territory before showing their hand and then release them to fly forward like an arrow. They burn boats and break cookware. Like a shepherd prodding sheep, they drive their soldiers every which way and none know where they are ultimately going.

The business of a General is to bring the enemy into danger. Nine different measures are suited to the nine different grounds.

The basic laws of human nature and the use of offensive or defensive strategies and tactics are studied. When invading enemy territory, the general principle is that penetrating deeply brings cohesion and shortly brings dispersion.

When home country is left and the army is taken to neighboring ones, this is contentious ground. When there is communication on all four sides, this is intersecting ground. When there is deep penetration into enemy country, this is serious ground. When penetration is slight, it is facile ground. When the enemy's stronghold is at the rear and narrow passes are in front, this is hemmed-in ground. When there is completely no place of refuge, this is desperate ground.

On dispersive ground, soldiers are inspired with a unity of purpose. On facile ground, there is a close connection between all parts of the army. On contentious ground, the rear is hurried to. On open ground, vigilance is kept on defenses. On intersecting ground, allies are consolidated. On serious ground, a continuous flow of supplies is ensured. On difficult ground, quickly pushing along the road is necessary. On hemmed-in ground, ways of retreat are blocked. On desperate ground, the hopelessness of surviving and the need to fight to the death are acknowledged.

It is a soldier's disposition to resist when surrounded, to fight hard when it cannot be avoided and to promptly obey commands when in danger. Alliances with neighboring States cannot be made without knowing their designs. Able leading and maneuvering of an army requires knowing the advantages and pitfalls of the country and its mountains, forests, marshes and swamps and making use of local guides. All of these things must be known in order to use terrain advantageously and be worthy of leadership.

Several principles to be followed by noteworthy military leaders are:

1) When attacking a powerful state, concentration of enemy forces are prevented by astounding them and discouraging allies

from joining forces with them.

2) Alliances are not made with just any State nor is authority and power over them used.

3) Secret designs are carried out that keep the enemy in awe and enable the capturing of cities and overthrowing of States.

4) Rewards are given and orders are issued without considering previous rules or arrangements and the whole army is treated, handled and advanced as one soldier.

5) Soldiers are presented with strategic tactics without knowing the design behind them.

6) Strategic advantages are made known to soldiers and tactical disadvantages are not.

7) Placing the army in peril and deadly terrain, it will safely survive and live because it is capable of overcoming defeat and fighting for victory.

8) Success in warfare is accomplished by cautiously and deceptively accommodating to the purposes of the enemy while directing forces to vulnerable spots.

9) Sheer cunning and continually keeping forces on the enemy will eventually succeed in killing the Commanding General. This is using the enemy army to strengthen one's own.

10) On the day of taking command, borders are blocked, passports are destroyed and the passage of emissaries is stopped. Calculation and plan rehearsal in the ancestral temple will control and assess the situation prior to executing plans and strategies.

11) When the enemy leaves a door open, rush in. Forestall the enemy by quickly occupying some place that is valued by them and time the enemy's arrival on the ground.

12) Maintain discipline, learn about and adjust to the enemy's circumstances until able to fight a decisive battle.

13. At first, exhibit the shyness of a young maiden until the enemy has an opening and then spring forward with the speed of a darting rabbit and it will be too late for the enemy to resist.

THE SYNOPSIS

In the *Art of War*, nine grounds, terrains or situations are recognized:

1) Dispersive ground – when the enemy is fighting in its own territory. Do not fight.

2) Facile ground – when the army has entered into enemy territory. Do not stop.

3) Contentious ground – possession is advantageous to either side. Do not attack.

4) Open ground – both sides have freedom of movement. Do not become separated.

5) Intersecting ground – three adjacent States when won control an empire. Join allies.

6) Serious ground – when the army penetrates deep into enemy country. Plunder.

7) Difficult ground – terrain that is difficult to traverse, Continue marching steadily.

8) Hemmed-in ground – land with narrow gorges and tortuous exits. Resort to strategy.

9) Desperate ground – immediate fighting is the only way to escape death. Fight!

The skillful General separates the enemy front and rear lines to disrupt coordination.

Enemy soldiers are kept disordered. Forward moves are only made when advantageous.

When an array of enemy soldiers marches to attack, something they value is taken.

Speed is the essence of war. Advantage is taken of the enemy's unpreparedness.

The army's way is made by unexpected routes and unguarded places are attacked.

The following principles are to be observed by an invading army:

1) Further penetration into enemy territory results in greater troop solidity.

2) Forays are made into fertile country in order to supply the army with food.

3) Soldiers are not overly taxed and their well-being and morale are monitored.

4) Energy is concentrated and conserved. Unpredictable strategies are devised.

5) Soldiers are put in inescapable situations and become fearless faithful fighters.

6) Omens are not made and superstitious doubts are dispelled to avoid fears.

On battle day, soldiers may be anxious but once in battle they are courageous.

Enemy soldiers caught in the same dire straits become supportive and assisting.

An army is managed by setting one uniform standard of courage that is shared.

Skillful Generals conduct an army as though leading a single soldier by the hand.

They are quiet to maintain secrecy and are upright and fair to maintain order.

They mystify officers and soldiers with misinformation to keep them unknowing.

They keep the enemy ignorant by changing plans and actions and prevent them from anticipating purposes by changing campsites and taking circuitous routes.

The purpose of the General is to bring the enemy army into a dangerous position.

Laws of human nature and the use of offensive and defensive strategies are studied.

When invading enemy land, deep penetration brings cohesion, short brings dispersion.

Nine different measures are suited to nine different grounds or situations:

1) Dispersive – the army is fighting in homeland. Soldiers inspired by a unity of purpose.

2) Facile – slight incursion into enemy land. Close connection between parts of the army.

3) Contentious – homeland is left to go to enemy land. The rear is hurried to.

4) Open – both sides have free access to the land. Vigilance is kept on defenses.

5) Intersecting – communication is on all four sides. Allies are consolidated.

6) Serious – deep incursion into enemy land. A continuous supply-flow is needed.

7) Difficult – precipitous and difficult to traverse. Pushing along the road is necessary.

8) Hemmed-in – enemy at rear and narrow passes in front. Ways of retreat are blocked.

9) Desperate – no place for refuge. The hopelessness of survival is acknowledged.

Soldiers naturally resist when surrounded and fight hard when it cannot be avoided.

Alliances with neighboring States cannot be made without knowing their designs.

Able leading of an army requires knowledge of the land and the use of local guides.

Noteworthy military leaders follow several principles:

1) When attacking a powerful enemy, astound them and discourage allies from joining.

2) Alliances are not made with just any or every State and their power is not usurping.

3) Secret designs are carried out to keep the enemy in awe and enable overthrowing.

4) Rewards are given without considering past rules and soldiers are treated equally.

5) Soldiers are given the tactical strategy without being told of the design behind it.

6) Strategic advantages are shared with soldiers and tactical disadvantages are not.

7) Placing an army in deadly peril, it will survive and live due to fighting for survival.

8) Warfare success is accomplished by deceptively accommodating to enemy purposes.

9) Cunning and keeping military forces on the enemy will result in killing the General.

10) On command day, borders are sealed, passports are destroyed and envoys blocked.

Calculations and rehearsals will control the situation prior to actually executing plans.

11) When the enemy has an opening, rush in. Forestall the enemy by taking valuables.

12) Maintain discipline and learn about the enemy until able to fight a decisive battle.

13) Initially, be shy like a young maiden until the enemy has an opening and then spring forward with the speed of a darting rabbit and it will be too late for the enemy to resist.

THE COMMENTARY

In this chapter nine kinds and characteristics of ground or situations are identified, outlined and detailed along with corresponding positions to be taken or avoided and strategic tactics to be used or modified by our whole ego-Self that influence and determine effectively or ineffectively dealing with and successfully or unsuccessfully resolving conflicts with opposing whole ego-Selves, e.g., knowing the circumstances and 'the lay of the land' of conflicts are essential for effectively dealing with them and for successfully resolving them

The nine grounds or situations are:

1) Dispersive – when our whole ego-Self is conflicting with

an opposing whole ego-Self in its own surroundings, e.g., when meeting at their own or usual place. Do not fight with the opposing whole ego-Self.

2) Facile ground – when our whole ego-Self has entered the surroundings of an opposing whole ego-Self, e.g., when approaching their own or usual place. Do not linger in the unfamiliar surroundings.

3) Contentious ground – when possession is by both our whole ego-Self and an opposing whole ego-Self and is mutually advantageous, e.g., when meeting at a mutually chosen and suitable place. Do not advance or attack in this situation and wait for moves made by the opposing whole ego-Self.

4) Open ground – when both our whole ego-Self and an opposing whole ego-Self have freedom of movement, e.g., when meeting in a park. Stay connected with the opposing whole ego-Self.

5) Intersecting – when the first arrival commands adjacent surroundings, join with allies, e.g., reach out to supportive friends.

6) Serious – when our whole ego-Self has penetrated deeply into the surroundings of an opposing whole ego-Self. Adopt useful methods and techniques from the opposing whole ego-Self, e.g., learn and use circumstances and 'the lay of the land'.

7) Difficult – when the surroundings and ways of conflict are difficult to traverse. Move through them swiftly and steadily, e.g., quickly move away from difficult issues.

8) Hemmed-In – when there are tight spots in ongoing conflicts and complicated ways of exiting them. Devise and resort to effective strategic tactics, e.g., find ways to move out of tight spots.

9) Desperate – when direct fighting with an opposing whole ego-Self is the only way to not be defeated and destroyed. Fight as strongly and as fully as you can, e.g., be the most empowered and assertive as possible.

Our skillful, effective and successful executive ego-self knows how to divide and weaken an opposing whole ego-Self by disorganizing and disordering the integration, interaction, coordination and support of its subordinate ego-selves, e.g., by engaging and defeating one or more of the opponent's stronger subpersonalities and thus fragmenting and weakening an opposing whole ego-Self.

Forward movements are only made against an opposing whole ego-Self when they are advantageous, e.g., areas of strength and weakness are assessed prior to advancing. When the subordinate ego-selves of an opposing whole ego-Self is well-organized and well-ordered, something of value is taken from them, e.g., challenging a favored position or defeating an attached belief of a subpersonality.

Speed is an essential determining factor in skillfully, effectively and successfully dealing with and resolving conflicts. Strategic tactical maneuvers are made swiftly, e.g., at times quickly interrupting an opponent's arguments during a conflict. Advantage is taken of an opposing whole ego-Self's lack of preparation by surprise and unpredictable moves, e.g., by catching an opposing whole ego-Self completely off-guard during a conflict with some strong, spontaneous and abrupt pointed expression.

When our whole ego-Self moves away from defensive and protective strategic tactics to offensive and advancing ones, the following are relevant principles:

1) Deeper penetration into an opposing whole ego-Self's territory results in greater solidity of our subordinate ego-selves, e.g., increasing the sense of the unity of our whole ego-Self through incisive expressions.

2) Forays into the resourceful areas of an opposing whole ego-Self and its subordinate ego-selves, e.g., using some of their strategic tactics and forward movements to overthrow its various subpersonalities as in ju-jitsu, judo and aikido.

3) Carefully monitoring the viability, vitality and activity

levels of our subordinate ego-selves and not overtaxing them during conflicts, e.g., enlisting and appropriately using numerous relevant subpersonalities and not just a select and preferred few.

4) Concentrating and conserving energy and strength and devising unpredictable strategies, e.g., interacting with an opposing whole ego-Self minimally, tactically, succinctly and emphatically.

5) Placing our subordinate ego-selves in positions where they naturally need to, and are fearlessly able to, use their most strength, e.g., appropriately using certain of our courageous subpersonalities in critical circumstances and during urgent conflicted interactions with an opposing whole ego-Self.

6) Prohibiting the use of omens and superstitions to assess the progress and to predict the outcomes of conflict, e.g., not being distracted from making real, accurate, definite and undoubted evaluations based upon the actual conduct of conflicts and ways they are going.

Before directly engaging in conflicts, our subordinate ego-selves may be anxious and afraid but once actively involved in them, they demonstrate the ferocity of a legendary snake that quickly counterattacks with head and tail when either, both or its middle are attacked, e,g., enlisting and allowing various of our subpersonalities to become actively involved and quickly come to aid each other when individual ones are attacked in conflicts with an opposing whole ego-Self.

In intense struggles and conflicts between opposing whole ego-Selves that threaten the viability and integrity of both, opposing subordinate ego-selves can form alliances and come to each other's aid in order to prevent that from occurring, e.g., a powerful negative event or dire circumstance equally impacting individual opponents or enemies may compel them to transcend their conflict or warring and to relationally join together and whereby a strong subpersonality of our whole ego-Self may rescue a weaker one of an opposing whole ego-Self.

In dealing with and resolving conflicts, it is insufficient to simply secure the means of engagement and it is important to unify our subordinate ego-selves by our executive ego-self; setting one uniform standard of courage for all of them to collectively reach, maintain, share and enact in conflicts, e.g., well-founded and well-ordered presentations of arguments need to be backed up with the force of solidly integrated subpersonalities.

Our skillful, effective and successful executive ego-self:

1) leads by making the most of both the strengths and weaknesses of our subordinate ego-selves, e.g., understanding the unique natures and characteristics and using the individual talents and skills of our subpersonalities.

2) connects with numerous of our subordinate ego-selves as if they are a single unified body, e.g., our subpersonalities are harmoniously unified with one another and constitute and sustain the integrity wholeness and power of the our whole ego-Self.

3) is calm and quiet to ensure secrecy, upright and fair to maintain order and may mystify our subordinate ego-selves but only for purposes of keeping them completely focused on the matter at hand and their essential and necessary roles in conflicts, e.g., our subpersonalities are not distracted, deviated or deterred from sharing and enacting a united purpose when being used in dealing with and resolving conflicts.

Our skillful, effective and successful executive ego-self:

1) knows the correct use of grounds, terrains, conditions, circumstances and situations.

2) varies strategic tactics and maneuvers in accord with unique and changing situations.

3) keeps opposing whole ego-Selves in the dark by revising plans, strategies and tactics and changing positions, dispositions and expectations, e.g., being familiar with differing kinds of conflicts, flexibly modifying various interactions and shifting the focus and direction of conflicting disagreements, disputes, arguments and fights with an opposing whole ego-Self.

At critical strategic moments in conflicts, our executive ego-self concentrates, releases and spreads out the synergistic power and overwhelming force of our united subordinate ego-selves against an opposing whole ego-Self, e.g., by consolidating, focusing, directing and radiating the energies of our subpersonalities with laser-like intensity and precision.

The business of our executive ego-self is to bring opposing whole ego-Selves into positions of danger and vulnerability and nine different measures are suited to the nine different types of grounds and situations.

The basic laws of human nature and the use of defensive and offensive strategic tactics are studied and used. The general principle when invading the territory of an opposing whole ego-Self is that a deeper penetration brings about the cohesion of our subordinate ego-selves and a shorter one brings about their dispersion, e.g., understanding human nature in general, moving into deeper areas of an opponent's unique nature and key conflicting issues with strong, relevant, focused and consolidated subpersonalities.

The nine different measures for the nine different conditions and situations are:

1) Dispersed – our whole ego-Self is inspired by the unity of purpose of safeguarding itself, e.g. our executive ego-self consciously pulls together and unites our subpersonalities.

2) Facile – slight penetration into opposing territory. Our subordinate ego-selves stay connected, e.g., our executive ego-self consciously keeps our subpersonalities from scattering.

3) Contentious – our whole ego-Self goes to the opposition's territory. The rear is hurried to, e.g., our executive ego-self relies upon the background support of our subpersonalities.

4) Open – vigilance is kept on the opposition and offenses and defenses are mutually used, e.g., our executive ego-self needs to closely observe the offensive moves of the opposition.

5) Intersecting – four-sided interconnection of territories. Allies are made and consolidated, e.g., our executive ego-self

enlists the support of closely adjacent subpersonalities.

6) Serious – deep penetration into opposing territory. A continuous flow of supplies is made, e.g., the supportive resources of all relevant subpersonalities are enlisted and put into play.

7) Difficult – territory is difficult to traverse. Quickly and steadily pushing along is necessary, e.g., our executive ego-self and subpersonalities keep on navigating through the conflict.

8) Hemmed-In – the opposition is at the rear and narrow passes are in the front. Ways of retreat are blocked by the opposition, e.g., an opponent stands in front of an exit door.

9) Desperate - there is completely no place of refuge. The hopelessness of survival is acknowledged and there is a fight to the finish, e.g., our executive ego-self and the subpersonalities of our whole ego-Self do their best and battle their hardest in an all-out fight with opposing ones.

It is the disposition of our subordinate ego-selves to resist when surrounded by an opposing whole ego-Self and to fight hard when conflict cannot be avoided and commanded to do so by our executive ego-self, e.g., our subpersonalities naturally resist external controls, follow internal instincts and fight hard against strongly opposing and actively conflicting whole ego-Selves.

Potential allies that might possibly be used to support our whole ego-Self involved in conflict are assessed as to their motives before forming alliances, e.g., friends who are needed for active support in dealing with and resolving conflicts need to be free of their personal needs and investments in order to be useful and effective.

Competent leading and maneuvering of our subordinate ego-selves in navigating the process of conflicts require that our executive ego-self knows the advantages and the disadvantages and the risks and dangers of engaging in certain conflict situations and enlists the guidance of other whole ego-Selves who are familiar with them, e.g., navigating the landscape and terrain of conflicts is benefited by the assistance of other human beings who have experienced similar ones.

Our wise, effective and successful executive ego-self adheres to the following principles:

1) Astounding advances on powerful opponents scatter their forces and discourage allies, e.g., a dramatic all-out advance on opponents may induce awe, fear and disorganization.

2) Alliances are made with selected allies and their power is not used to replace our own, e.g., supportive allies are useful in confrontations with opponents but not disempowering.

3) Enacted secret plans keep opponents in awe and enable overthrowing and capturing them, e.g., opponents can be mesmerized, startled and overwhelmed by unfolding covert plans.

4) Rewards are not fixed and are given to our subordinate ego-selves specific to conflicts and the entire number of our varied subordinate egos are handled and advanced as one whole, e.g., a pre-arranged structure of rewards for subpersonalities is modified as needed.

5) our subordinate ego-selves are told of strategic tactics but not the designs behind them, e.g., subpersonalities are not distracted, deviated and deterred by focusing upon them.

6) Strategic advantages are made known to subordinate ego-selves but not disadvantages, e.g., our subpersonalities are positively motivated rather than negatively preoccupied.

7) Our subordinate ego-selves placed in dire straits and deadly peril will fight hard and survive, e.g., the maximum strength of subpersonalities enables them to defeat opponents.

8) Success in warfare depends upon cautiously and deceptively accommodating to the intents of an opposing whole ego-Self while directing forces to advance against its vulnerable spots, e.g., appearing to acquiesce to an opponent while simultaneously attacking weaknesses.

9) Sheer cunning and continually advancing on an opposing whole ego-Self eventually succeeds in subduing its executive ego-self and using the opposing whole ego-Self to strengthen

one's own, e.g., consistently and persistently out-foxing opponents ultimately leads to their defeat.

10) On the day of beginning conflict, borders are closed, passports are destroyed, envoys are banned and plans are calculated and assessed in the ancestral temple prior to execution, e.g., at the beginning of a conflict, ego-boundaries are secured in the heart-mind of our executive ego-self of our whole ego-Self and plans are reviewed, contemplated and rehearsed in a sacred place.

11) Rush in when an opposing whole ego-Self leaves an opening. Forestall an opposing whole ego-Self by quickly occupying a valued place and await and time its arrival on the conflict ground, e.g., the open and valued places of opponents disadvantages them in conflicts.

12) Keep discipline, discern and adjust to an opposing whole ego-Self until able to fight a decisive battle, e.g., maintaining the integrity of our whole ego-Self and adjusting to the positioning and movements of an opponent and not engaging until defeat is possible.

13) In conflicts, initially exhibit the shyness of a young maiden until the opposing whole ego-Self has an opening and then spring forward with the speed of a darting rabbit and it will be too late for it to resist and the way to its ultimate defeat opens, e.g., initially seducing an opponent results in an opening that is vulnerable to a quick attack and ultimate defeat.

Win by knowing situations,
employing strategic variations
and effective tactical creations.

Your Reflections

CHAPTER TWELVE
FIRE ATTACKS

火　　　攻

HUO/HUO
Fire/Flames

KUNG/GONG
Attack/Assault

HUO/HUO – fever/burn

KUNG/GONG – art/work
+
P'u/Pu – tap/rap/knock

THE TEXT

Sun Tzu says:

There are five ways of attacking with fire:

1) Setting fire to the camp of soldiers.

2) Setting fire to storehouses and provisions.

3) Setting fire to supply wagons and equipment.

4) Setting fire to arsenals and granaries.

5) Setting fires along communication and supply lines and reinforcement routes.

To carry out a fiery attack, material for use in fires should be available and ready. The correct season for fires is when the weather is very dry and grasses are high and the special days are when the moon is in relation to certain constellations and strong winds arise.

In fire attacks, preparations need to be made to meet and follow up with five possible developments:

1) When fire breaks out inside the enemy camp, respond

quickly with an attack from outside.

2) When there is an outbreak of fire but enemy soldiers are quiet, wait and do not attack.

3) When fires have peaked, follow up with an attack if practicable, otherwise remain in place.

4) When attacking with fire from outside, do not wait for it to break out inside and attack at a favorable time.

5) When starting a fire, stay upwind of it and do not attack from downwind.

A long lasting daytime wind will end by night. All armies must know these five developments connected with fire and their conduct and timing.

Using fire to assist attacks is intelligent. Using water, accesses strength. By means of water, an enemy can be isolated but not pillaged.

Attempting to succeed in attacks and win battles without following through is a waste of time. Therefore, an enlightened Ruler and a wise Commanding General cultivate resources; consider, refine and implement strategic plans well ahead of time and reward merit.

They do not move unless an advantage is seen. They do not use troops unless there is something to be gained. They do not fight when there is no danger. A Ruler or Commanding General should not mobilize its army out of anger or spite, field their troops simply for self-gratification or fight battles simply out of wounded vanity.

When it is advantageous and beneficial, an offensive and aggressive forward move is made, otherwise remain in place. Anger and spite may, in time, change into happiness but a State once destroyed can never be restored nor can the dead ever be revived.

Therefore, an enlightened Ruler and a wise Commanding General are alert and attentive to this and are prudent and cautious. This is the Way to keep the State secure and peaceful and to preserve an army intact and whole.

THE SYNOPSIS

The five ways of attacking by fire are:

1) Setting fire to the camp of soldiers, 2) setting fire to storehouses and provisions, 3) setting fire to supply wagons and equipment, 4) setting fire to arsenals and granaries and 5) setting fires along communication and supply lines and reinforcement routes.

Material for fires need to be ready, grasses should be high and the weather dry and windy.

Preparations need to be made to meet and to follow-up with the following situations:

1) Quickly respond to fire breaking out in the enemy camp and follow up with an attack.

2) When enemy soldiers are quiet during a fiery outbreak, wait and do not attack.

3) When flames have reached their height, follow up with an attack if practicable.

4) When attack fire is outside, don't wait until it is inside and attack when favorable.

5) When starting a fire, always stay upwind of it and don't attack from downwind.

Fires must be accurately timed. Daytime winds lasting long will stop by night.

Fire attacks show intelligence. Water attacks access strength and can isolate enemies.

Succeeding in attacks and winning battles need good following up to not waste time.

Enlightened Rulers cultivate resources and wise Generals refine strategies ahead of time.

They don't move until there is an advantage and don't use troops unless there some gain.

They don't mobilize an army or fight out of anger or spite or for self-gratification or vanity.

Aggressive forward moves are made only when advantageous otherwise they stay in place.

Anger and spite may in time turn into happiness and wounded vanity may in time heal but a State once destroyed can never be restored nor can the dead ever be revived.

Thus an enlightened Ruler and a wise Commanding General are alert and attentive to this and are prudent and cautious with respect to war and warfare. This is the Way the State is kept secure and peaceful and an army is kept whole and intact.

THE COMMENTARY

Resorting to and making inflammatory attacks on an opposing whole ego-Self is an intense, strategic and tactical way of provoking and disrupting it; diverting and preoccupying it with protecting and saving itself and rescuing subordinate ego-selves; extinguishing fiery and heated fights; discouraging it from engaging in battling and accelerating the quick, decisive and final ending of conflict.

The five types of attacks with fire are:

1) Attacks on the encampment of the executive ego-self and the subpersonalities of an opposing whole ego-Self, e.g., igniting and inflaming an opposing whole ego-Self by fiery rhetoric and heated confrontation.

2) Attacks on the storehouses and provisions of the executive ego-self and subpersonalities of an opposing whole ego-Self, e.g., defusing the necessary kindling and preparations for conflicts by an opposing whole ego-Self.

3) Attacks on baggage wagons and equipment, e.g., disarming the personal ego-skills and abilities that an opposing whole ego-Self carries into and uses for conflicts.

4) Attacks on granaries and arsenals, e.g., dismantling the natural dispositions, ego-weaponry and fighting skills that an opposing whole ego-Self brings into conflict.

5) Attacks on communication and supply lines and reinforcement routes, e.g., disabling the transmission of information between the executive ego-self and subpersonalities of an opposing whole ego-Self, the need-satisfying ego-maintainers of an opposing whole ego-Self and its additional ego-strengtheners.

Attacks by fire require available combustible material and suitable conditions, e.g., intensely passionate burning desires, ardent feelings and fervently blazing actions on the part of our executive ego-self and our subpersonalities of our whole ego-Self under optimal and perfectly conducive conditions that will ensure defeating an opposing whole ego-Self.

For specific fire attacks, certain actions need to be taken and their developments followed up with which are:

1) When inflammation kindles, ignites and catches fire within and inflames an opposing whole ego-Self; our whole ego-Self can act quickly with an advance or attack from outside, e.g., capitalize on an opponent's inflammation.

2) When, in the heat of a fiery situation, an opposing whole ego-Self remains quiet; our ego-Self waits and does not advance or attack, e.g., allow an opponent to do a slow burn.

3) When the intensity of a fiery situation has peaked, it is followed up by our whole ego-Self with a brilliant advance or attack on the opposing whole ego-Self if and when feasible, e.g., capitalize on a blazing and highly well-lit situation.

4) When the outer appearance, demeanor and composure of an opposing whole ego-Self has been inflamed but has not gone deeper inside, an advance or attack can be made by our whole ego-Self at a favorable time, e.g., capitalize on the surface irritations and annoyances of an opponent.

5) When inflaming an opposing whole ego-Self, our ego-Self remains upwind and distant and does not advance or attack from downwind and close so as to not be ignited itself, e.g., stand back at some distance from an opponent.

Our wise, effective and successful executive ego-self knows, adapts to and uses these five actions and developments of fiery advances, attacks and follow-ups and times them in accord with suitable incendiary conditions, in appropriate situations and under advantageous circumstances.

Using fiery advances and attacks demonstrates the intelligence and employs the strategic tactical proficiencies and competencies of our executive ego-self, e.g., its offensive energy, power, strengths and abilities.

Watery advances and attacks can also be made with the intent of demonstrating the force and strengths of our whole ego-Self and of isolating and dispersing an opposing whole ego-Self, e.g., hosing, watering and dampening down an angry mob of protesting subordinate ego-selves of an opposing whole ego-Self.

Our wise executive ego-self of our whole ego-Self is effective and successful in defeating an opposing whole ego-Self and resolving conflicts by having an energetic and enterprising spirit that is ready to take the initiative and to pre-plan, undertake and implement bold, purposeful and daring strategic tactical actions in difficult, complicated and risky situations without wasting time, energy and ego-self resources, e.g., our whole ego-Self engages in and handles conflicts with an opposing whole ego-Self actively, boldly, swiftly and economically.

Our wise, effective and successful executive ego-self does not act unless an advantage is seen and does not utilize our subpersonalities unless a benefit is to be gained. It does not fight when there is no danger. It does not mobilize our subpersonalities to offensively aggress against, engage with and battle an opposing whole ego-self out of anger, self-gratification and aggrandizement or wounded vanity, e.g., our whole ego-Self calmly and selflessly engages advantageously and beneficially in conflicts with an opposing whole ego-Self.

Violent anger, deep hatred, flagrant grandiosity and flaming pride eventually may and can transform into calm, acceptance, self-dignity and self-respect; but the State of Tao and our True

Tao-Nature, once absolutely annihilated, can never be restored and human beings, once killed, can never be revived.

Therefore, our enlightened True Tao-Nature, our integrated whole ego-Self and our wise, compassionate and skillful executive ego-self treasure precious human beings and cherish blessed human life, are aware of and attentive about the destruction and devastation caused by conflict and war and are cautious, prudent and economical when forced into being involved them, e.g., opposing whole ego-Selves are regarded with dignity, respect, humaneness and compassion.

By subduing and defeating an opposing whole ego-Self, by successfully overcoming and resolving conflicts and by preventing and ending conflicts without fighting them; awakened human beings and enlightened leaders secure and safeguard the State of Tao, preserve and sustain our True Tao-Nature and create and insure the viability, dignity, integrity and vitality and the peacefulness, freedom, happiness and intimate community of our human being and that of our fellow human beings.

Closing Comment

This concludes Chapters Nine, Ten, Eleven and Twelve on non-personal relationships and conflict. Non-personal conflict is successfully resolved by, through, in and as the harmonious co-existing and cooperative interrelationship of our whole ego-Self with the nature, characteristics, conditions, situations and requirements of our environment. As such, this preserves the original unity, integrity, peacefulness and pacifism of our True Tao-Nature and Tao-State within our awakened, enlightened and liberated consciousness, conscious awareness and conscious experience of our non-personal relationships.

An incendiary attack
involves military knack
and holds the enemy back.

Integrating ego and environment
without distortion or detriment.
The ego-self's great experiment
conducted with deep sentiment.
The Soul free from impediment
Spirit harmonizing the firmament.

Some Examples

Since the commentaries are also generalized abstractions of non-personal relationships, it may be helpful to provide concrete examples of the relationships between our executive ego-self/I, subordinate ego-selves/me-s and whole ego-Self/human being and the situational conditions of conflicts.

1. Our executive ego-self/I may identify with and employ its strategic sub-personality/me in understanding the terrain of a conflict and a suitable positioning within it in order to not endanger our whole ego-Self/human being.

2. Our executive ego-self/I may identify with and employ its trickster sub-personality/me to outfox an opposing and conflicting whole ego-Self and to successfully end a conflict without a direct confrontation.

YOUR REFLECTIONS

TRANSPERSONAL RELATIONSHIPS

GO-BETWEENS

CHAPTER THIRTEEN
USING
GO BETWEENS

用　　　間

YUNG/YONG
Use/Employ/
To hit the center

CHIEN/JIAN
Between/Among/
A space between

YUNG/YONG – the use/means
+
CHUNG/ZHONG – center/
middle

MEN/MEN – gate/sect
+
JIH/RI – sun/day

THE TEXT

SUN TZU SAYS:

Raising large armies and marching them great distances entails heavy losses for the people and depletes the resources of the State. The daily expenditures will amount to thousands of ounces of gold or silver. There will be commotion at home and abroad and people will fall down exhausted on roadways. The lives and work of hundreds of thousands of people and families will be disrupted, impaired and impeded.

Enemy armies may face each other for years striving for victories that are decided in a single day. This being so, remaining ignorant of the enemy's condition simply because of being unwilling to outlay a hundred ounces of gold or silver for wages and honors for intelligence gathering is the height of inhumanity.

One who acts like this is no help to a Ruler, no leader of soldiers and no master of victory.

Therefore, what enables an enlightened Ruler and a wise Commanding General to move and conquer and to achieve that which is beyond the reach of ordinary human beings is foreknowledge.

Such foreknowledge cannot be obtained from spirits, ghosts and astrological configurations or inductively derived from experience or deductively derived through calculation. Knowledge of the enemy's situation, condition and dispositions can only be obtained from human beings acting as go-betweens, spies and secret espionage agents.

The five classes of spies are: 1) Local spies, 2) Inside spies, 3) Converted spies, 4) Doomed spies and 5) Surviving spies. When all five of these are actively working, no one can discover the secret network that is a Ruler's most precious and valued resource.

Local spies are inhabitants of a given district. Inner spies are officials of the enemy. Converted spies are former enemy spies. Doomed spies are known, deceived and report back false information to the enemy. Surviving spies are bringing back intelligence about the enemy.

No military relationships are more intimate, and need to be confidentially maintained and liberally rewarded, than those with spies. In no other military business is greater secrecy preserved than in espionage.

Spies cannot be usefully employed without their intuitive discernment. They cannot be successfully managed without straightforwardness and benevolence. One's own subtle mental ingenuity is required to affirm the validity and truth of their reports. It is a most delicate and subtle art!

Spies are employed and useful for every kind of military business. When secret information is divulged by spies ahead of time, they are put to death along with to whomever secrets were told.

Whether the objective is to crush an army, storm a city or assassinate an individual; it is necessary to commission spies to find out the identities of, and detailed information about, all involved personnel. Enemy spies must be sought out, discovered, bribed, led away, comfortably housed and converted to double agents.

Through the intelligence brought by converted spies, local and inner spies can be acquired and employed, doomed spies can carry false information to the enemy and surviving spies can be appropriately used as planned for specific intelligence-gathering purposes.

The objective of all five varieties of spying is to covertly obtain knowledge from and about the enemy and all must be used. This knowledge is firstly obtained from converted spies and it is, therefore, essential to treat them with the utmost liberality and generosity. Therefore, only an enlightened Ruler and a wise Commanding General who use the greatest intelligence of spying and espionage will achieve the greatest success in warfare.

Relying upon spies and secret spying operations is a most essential factor in the *Art of War*, because upon them depends an army's every action.

THE SYNOPSIS

Raising large armies and marching the great distances entails heavy losses for the people and depletes the resources of the State. Daily expenditures will amount to thousands of ounces of gold or silver. There will be commotion at home and abroad and people will fall down exhausted on roadways. The lives and work of hundreds of thousands of people and families will be disrupted, impaired and impeded.

Enemy armies may face each other for years striving for victories that are decided in a day.

Being ignorant of the enemy's condition because of an unwillingness to fund intelligence is the height of inhumanity.

Acting like this is no help to a Ruler, no leader of of soldiers and no master of victory.

What enables an enlightened Ruler and a wise Commanding General to move and to conquer and to achieve what is beyond the reach of ordinary human beings is foreknowledge. Such foreknowledge cannot be obtained from spirits, ghosts or astrological configurations or inductively derived from experience or deductively derived from calculation.

Knowledge of the enemy's situation, condition and dispositions can only be obtained from human beings acting as go-betweens, spies and secret espionage agents. There are five classes of spies. When all five of these are actively working, no one can discover the secret network that is the ruler's most precious and valued resource.

1) Local spies – are inhabitants of local districts of the State.

2) Inside spies – are officials of the enemy State.

3) Converted spies – are former enemy spies or double agents.

4) Doomed spies – are known, deceived and report back false information to the enemy.

5) Surviving spies - bring back information and intelligence from the enemy.

No military relationships are more intimate and need to be more confidentially maintained and liberally rewarded than those with spies. If secrets are disclosed ahead of time, both the spy and the recipient are executed.

Spies are useful because of their intuitive discernment and are successfully managed with straightforwardness and kindness. Subtle ingenuity is required to assess and affirm the validity and truth of their reports. A most delicate and subtle art!

Spies are useful for every kind of military business. Whether the objective is to crush an army, to storm a city or to assassinate an individual; it is necessary for espionage agents to find out the identities of, and to gather information on, all personnel involved.

Enemy spies must be sought out, discovered, bribed, induced

to defect, comfortably housed and converted to double agents.

Through the intelligence brought by converted spies, local and inside spies can be recruited, doomed spies can bring false information to the enemy and surviving spies can be employed for specific planned intelligence-gathering purposes.

Only an enlightened Ruler and a wise Commanding General who use the greatest intelligence of spying and espionage will achieve the greatest success in warfare. Relying upon spies and secret spying operations is a most essential factor in the *Art of War*, because upon them depends an army's every action.

THE COMMENTARY

The prefix trans- is defined as 'across, by, through, beyond' and 'so as to change'. This final chapter of *The Art of War* goes beyond focusing upon the intrapersonal, interpersonal and non-personal conflicts of the previous twelve chapters.

It addresses the transpersonal resolution of conflict 'across and between' whole ego-Selves and opposing whole ego-Selves 'by and through' the use of intermediaries, spies and double espionage agents who spread false information to opposing whole ego-Selves and who obtain true intelligence about them 'so as to change' the nature and course of conflict from possible or probable defeat to real and actual victory.

The disrupting of the Tao-State, the displacing of our True Tao-Nature, the depletion of the resources of our whole ego-Selves, the destruction of the viability of our executive ego-self and our subpersonalities and the impoverishment and disruption of human beings and lives; are reiterated and re-emphasized.

Those in power positions who fail to administratively support and economically fund the use of spies, espionage operations and intelligence gathering in dealing with and resolving conflicts exhibit the extreme height of inhumanity and are no Rulers, leaders, masters or victors in conflicts, e.g., it is false economy for our executive ego-self to not support the go-between activities of our subordinate ego-selves that save the energy resources and preserve the integrity of our whole ego-Self.

What differentiates and elevates our awakened and enlightened True Tao-Nature, our harmoniously integrated whole ego-Self and our wise, compassionate and skillful executive ego-self beyond the reach of unawakened and unenlightened human beings is foreknowledge, e.g., the knowledge provided by the harmonious interrelationships of our subordinate ego-selves as go-betweens interacting and communicating with each other ahead of time that enables our whole ego-Self to effectively deal with and to successfully resolve conflict with an opposing whole ego-Self.

And foreknowledge: 1) is not obtained from astrological signs and configurations, mediums and soothsayers and turtle shell and yarrow stalk divinations or 2) is not inductively derived from concrete experience or deductively derived from abstract calculation, e.g., oracular methods or rational thinking.

Foreknowledge is obtained through the use of human beings who serve as intermediary, go-betweens, spies and secret agents actively engaged in clandestine, misleading and deception-transmitting and surreptitious information and intelligence-gathering espionage operations and relationships, e.g., using our subpersonalities that are in touch with each other and aware of each other's activities and those of opposing whole ego-Selves and can communicate them to our executive ego-self.

Five kinds of go-betweens, intermediaries, spies and secret espionage agents are:

1) Local spies – are our subpersonalities requisitioned from within our whole ego-Self.

2) Inside spies – are subpersonalities enlisted from within an opposing whole ego-Self.

3) Converted spies – are former subpersonalities of an opposing whole ego-Self working for our whole ego-Self.

4) Doomed spies – are subpersonalities of our whole ego-Self

that are given false information to be passed on to an opposing whole ego-Self.

5) Surviving spies – are our subpersonalities that bring back intelligence from and about an opposing whole ego-Self.

When all five kinds of our go-between spies and subpersonalities are engaged in combined active spying operations, they are the undiscovered secret system that is a most precious treasure and most valuable resource of the Tao-State, our True Tao-Nature, our executive ego-self and the leaders of human beings, e.g., it is imperative that the activities of our subpersonality go-betweens be kept secret by our executive ego-self in order to enable our whole ego-Self to gain advantages in conflicts and to make pre-emptive strikes upon an opposing whole ego-Self.

No relationships and operations in effectively dealing with and successfully resolving conflicts are more essential, critical and intimate and need to be most secretly preserved, confidentially maintained, freely used and generously rewarded than those with and of our subpersonalities acting as go-between spies, e.g., information and intelligence gathering by our subordinate ego-selves is indispensable for learning about the plans, strategies and tactics of an opposing whole ego-Self.

Our subpersonality go-between spies and double agents who are discovered to have revealed secrets ahead of time to the opposition are removed from espionage activities along with those to whom the secrets were disclosed, e.g., they are subordinate ego-selves that are no longer focused upon by our excutive ego-self in the conscious espionage activities of our whole ego-Self.

Our subpersonality go-between spies:

1) are useful and indispensable for every kind of intrapersonal and interpersonal conflict.

2) need to be intuitively discerning and managed straightforwardly and compassionately.

3) require subtle ingenuity to affirm the validity and truth of their reports, e.g., subtle sensing is necessary for our executive

ego-self to discern the veracity of intelligence obtained by and from our subpersonality go-betweens.

Whether the objective in dealing and resolving conflicts is: 1) subduing and defeating an entire opposing whole ego-Self or 2) deposing and eliminating a single opposing executive ego-self or leader; relevant information and intelligence concerning any and all involvements, associations, connections and activities need to be thoroughly investigated, completely acquired and accurately communicated, e.g., our subpersonality go-betweens need to access comprehensive critical data on an opposing whole ego-Self or an opposing executive ego-self and communicate it to our executive ego-self in order to be useful.

The go-between subpersonality spies of the opposition need to be sought out, discovered, bribed, led away, comfortably integrated and converted to double agents. Through the intelligence brought by converted subpersonality spies and double agents; local and inner subpersonlity spies can be enlisted, doomed subpersonality spies can pass on false information and surviving subpersonality spies can be appropriately used, e.g., the espionage activities of our subpersonality go-betweens is a complex network and ongoing flurry of information and intelligence-gathering and communicating with our executive ego-self of our whole ego-Self.

The essential overall objective of using all five subpersonality go-between spies is: 1) to covertly obtain essential and critical information and intelligence from and about an opposing whole ego-Self and 2) to overtly provide misleading and deceptive misinformation and disinformation to an opposing whole ego-Self, both of which result in its defeat in conflicts with our whole ego-Self without having to directly and actively engage it in fighting, combats and war, e.g., the espionage work of our subpersonality go-betweens greatly decreases and can prevent the need for our whole ego-Self to engage in head-on all-out fighting with an opposing whole ego-Self.

Only our awakened and enlightened True Tao-Nature, our wise and compassionate whole ego-Self and our executive ego-self: 1) who use the greatest intelligence obtained by our subpersonality go-between spies and 2) who use their greatest spying and espionage operations will achieve the greatest positive results in effectively dealing with and successfully resolving conflicts, e.g., the great eyes and ears of great subpersonality go-betweens contribute to and insure great conflict resolution for our whole ego-Self.

Relying upon our subpersonality go-between spies and spying operations is a most essential and critical factor in dealing with and resolving: 1) intrapersonal conflicts between our subpersonalities vying for supremacy and dominance between them, 2) intrapersonal conflicts of our subpersonalities with our executive ego-self and vying for autonomy and sovereignty and 3) interpersonal conflicts with our whole ego-Self and an opposing whole ego-Self, e.g., espionage operations and the activities of our subpersonality go-betweens is essential for effectively dealing with and successfully resolving both intrapersonal and interpersonal conflicts.

For, in conflicts, it is upon the bilateral interactions with, and the communication between, our subordinate ego-self go-betweens and our executive ego-self and between our subordinate ego-selves that depend every action intrapersonally within our whole ego-Self and interpersonally between our whole ego-Self and an opposing whole ego-Self, e.g., the use of subpersonality go-betweens for gaining foreknowledge determines the advantageous plans, strategies, positions, dispositions, tactics and maneuvers of our whole ego-Self involved in conflict with an opposing whole ego-Self.

CLOSING COMMENT

This concludes the final Chapter Thirteen on the transpersonal relationships and intrapersonal, interpersonal and non-personal conflict. Transpersonal conflict resolution is accomplished by, through, in and as the harmonious co-existing and cooperative interrelationship of intermediary go-betweens of our subordinate ego-selves with our whole ego-Self *vis-à-vis* those of an opposing whole ego-Self. As such, this preserves the original unity, integrity, peacefulness and pacifism of our True Tao-Nature and Tao-State within our awakened, enlightened and liberated consciousness, conscious awareness and conscious experience of our intrapersonal, interpersonal and non-personal relationships.

Dependable and reliable spies
assuming their artful disguise,
strategically and tactically wise,
are a victorious general's eyes
and a surviving ruler's prize.

The all-encompassing transpersonal,
including the personal and interpersonal
and the non-personal and impersonal.
The middle Way between all beings.
The going between of Whole Self and other,
of executive selves as father and mother,
of subordinate selves as sister and brother.

SOME EXAMPLES

Since the commentaries are also generalized abstractions of transpersonal relationships, it may be helpful to provide concrete examples of the go-between relationships of our executive ego-self/I, subordinate ego-selves/me-s and whole ego-Self/human being and those of opposing and conflicting whole ego-Selves.

1) Our executive ego-self/I may identify with and use its loyal subpersonality/me to discover any of its other sub-personalities/me-s that may pose threats to the overall integrity of our whole ego-Self/human being.

2) Our executive ego-self/I may identify with and enlist its intuitive sub-personality/me to obtain foreknowledge about and to discern problematic motives and activities of the executive ego-self and sub-personalities of opposing and conflicting whole ego-Selves.

CONCLUSION [7]

This is the conclusion of a rendition of Sun Tzu's ancient classic strategic masterpiece, *The Art of War* but, unfortunately, not the conclusion of war itself. In the Prologue, I shared some of my history with matters of war, conflict, and peace that was prompted by recollection and reflection. Now in the Conclusion, I am sharing some final sentiments about matters of war, conflict and peace that are prompted by grief and compassion.

War; including both unjustified and so-called 'just wars'; appears to be a tragic, regrettable and undeniable inevitability of the human condition. And, as Master Sun notes, perhaps the best we can do is to fight and end unavoidable wars wisely, compassionately, skillfully and as quickly as possible with the least depleting, dissipating and draining of economic resources; the least damaging, destroying and demolishing of community property and the least depriving, disrupting and dying of human beings and human Souls.

The Tao/Way/Art of war is a philosophy and psychology of warfare that involves:

1. Preparation – assessments, calculations, plans and forecasting made ahead of time.

2. Economics – quick victories and avoiding the drawbacks of costly protracted campaigns.

3. Strategies – subdue without fighting and the rules, conduct and essentials in warfare.

4. Tactics – positioning, defensive and offensive, measuring and estimating probabilities.

5. Energy – direct and indirect methods and conservation, power, momentum of energy.

6. Strengths – strengths and weaknesses and mystery, secrecy,

subtlety and flexibility.

7. Fighting – strategic tactical maneuvers, deception, *esprit de corps*, study of conditions.

8. Situations – variations in strategic tactics appropriately used in changing circumstances.

9. Logistics – positioning in various kinds of terrain and signs of enemy army movements.

10. Terrain – advantages, dangers and positionings in various kinds of situational ground.

11. Grounds – kinds of grounds and situations and conduct within them, principles in war.

12. Fire – kinds of fire attacks and appropriate follow-ups and conduct of the army forces.

13. Spies – the kinds, use, managing of go-betweens for foreknowledge and intelligence.

General principles of the Tao/Way/Art of war are:

1. Acknowledging the serious nature of war and the need for thorough understanding.

Having accord between Tao/the Way, Heaven and Earth, the leaders and the people.

Warfare involves deception, attacking an enemy where unexpected and unprepared.

2. Ending wars quickly and avoiding the resources and lives lost in prolonged campaigns.

Using the enemy army's captured resources to reinforce and strengthen one's own.

Effective leadership is a decisive factor in maintaining the safety and peace of people.

3. Subduing an enemy and winning a war is achieved by not engaging or fighting them.

Knowing when and when not to fight and being prepared against an unprepared army.

Understanding the enemy and knowing oneself insures victory in wars and battles.

4. Successful armies place themselves beyond the possibility of being defeated in wars.

Balancing strong and offensive and weak and defensive positions and maneuvering.

Establishing the certainty of victory by conquering an enemy that is already defeated.

5. Using both conventional and planned and unconventional and surprise maneuvers.

Strategically configuring and positioning united troops to build energy momentum.

Making momentous and perfectly timed military decisions and offensive attacking.

6. Keeping enemy armies unaware of knowing from where and when attacks are made.

Attacking weak places in the enemy army defenses and evasion of their strong ones.

Being mysterious, secret, subtle, formless, invisible, united and flexibly adaptive.

7. Knowing how to turn danger and disadavantage into safety and calculated advantage.

Carefully evaluating, deliberating, planning all army maneuvers and strategic tactics.

Observing and considering the mood, energy and organization of the enemy army.

8. Not relying upon the chance of the enemy not arriving but being prepared for them.

Not relying upon the likelihood of the enemy not attacking but being unassailable.

Leadership not being reckless, cowardly, hot-tempered, moralistic and overly kind.

9. Encamping and positioning military in places and ways according to environments.

Observing signs of the movements, advances, oncoming attacks of enemy armies.

Observing the activities, status and conditions of the personnel of the enemy army.

10. Determining the various kinds of terrain and appropriate positions to take in them.

Knowing errors of a leader that lead to the disorganization and defeat of an army.

Regarding soldiers as one's beloved children and with both kindness and discipline.

11. Recognizing and understanding the ground, terrains and situations of battlefields.

Devising strategies and contriving tactics that are appropriate to different terrains.

Taking measures and positioning and deploying troops suited to different grounds.

12. Using fire attacks to preoccupy the enemy with extinguishing and surviving them.

Using fire attacks to discourage, dishearten, demoralize, dispirit enemy soldiers.

The leader maintaining an enterprising spirit and doesn't waste time and effort.

13. Using go-betweens, intermediaries, spies, espionage agents and double agents.

Using spies to obtain intelligence about and foreknowledge of enemy's plans.

The leader recognizing the critical importance of using spies to achieve victories.

In the presence and midst of wars happening in our sacred world and on our precious planet, as human beings and human Souls living treasured and cherished human lives; we can, in the same here-now moment; focus upon, affirm, actualize, enjoy and share peace within ourselves and peace between ourselves and fellow human beings as incarnated Spirit, kindred Souls, sisters and brothers, wayfaring companions and sojourning pilgrims.

As incarnated Spirit, we are embodying the Mystery, Miracle,

Marvels and Magnificence of human birth, being and growing; human life, living and experiencing and intimate human co-existing and interrelating, communing and union.

As kindred Souls, we are a body-Spirit and Spirit-body located in the center of space-time, Heaven-Earth, our universe, our planet and our natural, physical, environmental, social, preternatural and supernatural worlds.

As sisters and brothers, we are descendants, progeny and off-springs of Great Mother Nature who are: 1) constantly, consistently, continuously and continually sunrising in the twilight of dawning days and moonrising in the twilight of dusking nights and 2) openly blossoming blue-green in Spring, brightly shining yellow-orange in Summer, fully ripening golden-brown in Autumn and softly glowing silver-white in Winter.

As wayfaring companions, we are journeying together on our paths as spiritual friends.

As sojourning pilgrims, we are briefly dwelling in sacred places before passing through.

Wayfaring not warfaring.
Empathizing and caring.
Intimacy and pairing.
Loving and sharing.

Sojourning and learning.
Human Heart discerning.
Human Karma burning.
Human Soul returning.

YOUR REFLECTIONS

Epilogue

Unfortunately, conflicts and wars are rampant and escalating throughout our sacred world, precious planet, interwoven countries and intimate global community. The pervasiveness, kinds and ways of various wars exist in reality and are also reflected in metaphors of that reality.

Hot wars, proxy wars, shadow wars, cyber wars, cold wars. Culture, religious, ideological, identity, gender wars. Political, civil, border, economic, information wars. The war on drugs; the war on abortion; the war on wages, poverty and homelessness; the war on immigration.

The war on gun control, the war on crime, the war on police, the war on violence and abuse. The war on health care; the war on inflation, interest rates, banks. The war on terrorism, racism, sexism, ageism, classism and the war on any and all of the various other 'isms'.

The war on climate, the war on technology and AI. The war on gas and oil, the war on Big-Pharma, the war on monopolies and big box stores, the war on unions. The war on social media, the war on free speech and censorship, the war on on-line child pornography.

The war on mental health, autism and dementia; the war on cancer, heart disease, diabetes and ageing; the war on obesity, diets, salt, sugar, fats, carbs, and gluten; the war on statins, medications and supplements; the war on vaccinations; the war on smoking, the war between the sexes.

The war on AIDS and HIV; the war on COVID, viruses, bacteria and germs; the war on plastic and synthetics; the war on toxins, pollutants and contaminants; the war between the sexes. The war on you name it! We need to wage a war on wars!

Such is the value and message of Sun Tzu's *The Art of War* that presents plans, strategies, positions, tactics and maneuvers that quickly bring about the ending of wars and conflicts and preventing them from reaching cataclysmic and apocalyptic proportion. They must be carefully and fully investigated, explored, studied, learned and employed in effectively dealing with and successfully resolving conflicts and ending wars within and between human beings, groups, organizations, institutions, societies, countries and nations of the world.

The future humanness and humaneness and the survival and evolution of our humanity and human species, culture, society and world critically depend upon our doing so. Much can be learned from studying the causes, conduct and results of significant historical wars that have occurred in our world and understanding the long-standing and ingrained systemic conflicts;

1) that have historically occurred in certain geographical locations, countries and nations;

2) that have historically occurred between the particular ethnic groups inhabiting them;

3) that likely may involve some kind of 'karmic' factors for the lands, leaders and peoples.

4) that exist buried deep within the collective unconscious of current land inhabitants;

5) that are embodied and repeated by nationalistic 'leaders' and governments and

6) that influence and drive current conflicts and wars between those certain countries.

The dangers for, and risks to, safeguarding, preserving, maintaining and sustaining the collective peace, freedom and happiness of our humanity have escalated; given that the increased fire power of modern military and civilian armament and weaponry now far exceeds that of ancient China, when *The Art of War* was written, and that of Medieval warfare and historically later militarized conflicts, revolutions and wars.

Armed conflicts and wars can be and, are now being, conducted using satellite technology, electronic decoys, robotic vehicles and thermal imaging; bombers, fighter jets, assault helicopters, armed drones and stealth aircraft; rockets, anti-ballistic and short- and long-range guided missiles and laser and hypersonic weaponry; nuclear warheads and atomic and hydrogen bombs; nuclear-outfitted warships and submarines; tanks and armored vehicles; land mines, flame throwers, bazookas, mortars and hand grenades; machine guns, automatic pistols and assault rifles rather than catapults, cannons, spears, lances, bows and arrows, crossbows, swords, clubs, maces, single-shot pistols and front-loaded matchlock or flintlock muskets.

Also, there currently appears to be an intensifying, escalating, upsurging, surfacing and outbreaking of a world-wide groundswell of insecurities, anxieties, dangers, fears, suffering, pain and anger over long-standing autocratic rule, tensions, power-plays, disagreements, hostilities and conflicts between various leaders and governments of countries and nations (not necessarily the peoples!) of the Middle, Far and Near East; North Africa; Eastern and Western Europe and North, Central and South America.

1) Some international super-power and non-NATO nations have withdrawn from arms control, nuclear non-proliferation and nuclear missile test ban treaties and the nuclear arms race is accelerating among world super-powers.

2) More defensive and offensive military combat armament and personnel are being placed on stand-by, emergency and red alert in strategic locations throughout the world;

3) Some foreign embassy closures and ambassador and civilian evacuations are occurring in countries threatened by potentially active wars;

4) More powerful and deadly bombs and more accurate and longer-range hypersonic missiles with nuclear warheads are being manufactured, stockpiled and tested and

5) The technology of the war-machine is relentlessly continuing to grind on, to become more sophisticated and developed and to be the critical measure of a nation's power and security and global superiority and dominance.

Due to seemingly geometrically and exponentially progressing, rising and escalating global conflicts and domestic crimes, our United States of America is in danger of becoming and being an international military police force and a national civilian police state.

Hopefully, the current prevailing and devolving imbalance skewed in favor of negative, divided, fragmented, violent, conflicting and warring individuals, factions and nations will soon reach a maximum point, will raise our enlightened consciousness of, and spiritual commitment to, preserving and sustaining our human being and human life and will naturally be counter-acted and compensated for; corrected, balanced and stabilized by, and reversed to, being in favor of more evolved, positive, harmoniously unified and peaceful ones.

Personally, with passion and candor toward the ending of my own precious human being, cherished and treasured life on our Planet Earth; I am disheartened and disappointed by:

1) our religion and politics having little to do with Divinity, Spirit, Soul and Human Beings.

2) our illusional egos and deluded selves and our attachment to impermanent things.

3) our discriminating and segregating and our differences, disputes, conflicts and wars.

4) our inequality between races, nationalities, genders, beliefs and societal position.

5) our socially white supremacy, alpha male dominance and omega female inferiority.

6) our societal capitalism, corporatism, collectivism, commercialism and consumerism.

7) our addictions to money, drugs, material things, success, sex and entertainment.

8) our diseases, illnesses, sicknesses, disorders, poverty, indigence and homelessness.

9) our striving, stretching, straining, stressing, sweating, struggling and suffering.

10) our parasitic, predatory and exploitative relations and unethical business practices.

11) our governmental politics being clandestine, corrupt, fragmented and dysfunctional.

12) our apparent ignorance of the sacred nature of our incarnation as human beings.

13) our inability to humanely resolve disputes and conflicts without resorting to wars.

14) our failure to globalize a growing collective dissatisfaction with autocratic leaders.

15) our lack of a sufficiently awakened collective consciousness and commitment to peace.

16) our apparent opting to wield the love of power more than to share the power of love.

Transpersonally; with wisdom, compassion, generosity, grace and forgiveness; I realize and accept that ending the wars, curing the diseases, resolving the conflicts, healing the ills and preserving the sanctity, dignity and integrity and sustaining the beauty, decency and civility of our humanity in our human world at large will likely never completely occur.

But consequences of epidemic, cataclysmic and apocalyptic proportion and Armageddon-like possibility can be mitigated to a considerable degree and meaningful extent by the collective agreement, intention, dedication, commitment, energy, courage, creativity, effort and ability of a significant number, and possibly critical mass, of awakened, enlightened, liberated, real, true, kind, caring and loving human beings.

And interpersonally; I know that, at the very least, the most concrete and direct, and the most readily available and accessible, thing that we all can immediately do in the present space-time and here-now of our lives is to awaken to peace, to make peace and to be at peace;

1) intrapersonally between the many diverse sub-personalities within ourselves;

2) interpersonally with our fellow human beings, kindred human souls and Spirit;

3) non-personally (not impersonally!) with our environment and other species and

4) transpersonally with our original and universal inborn True Tao-Nature; the unique and radiant Soul and the infinite and eternal Spirit that we all absolutely and ultimately, essentially and fundamentally, purely and clearly, really and truly, deeply and fully and freely and openly; always have been and are, beyond real and unreal, good and evil, true and false, right and wrong, peace and war, life and death.

Lao-Tzu says: 'As compassionate human beings, we are bearing the inner shame of our country and being custodial guardians of our land. As wise human beings, we are enduring the outer misery of our country and being equitable stewards of our world.' *Tao Te Ching* Passage #78 (Rendered).

The purpose of government
is to ensure the safety, security,
health and well-being of people.

Protect, defend and preserve
yourself, family and country.

YOUR REFLECTIONS

Conflict & War, Harmony & Peace

Conflict & War	Harmony & Peace
Competitors	Counterparts
Opponents	Complements
Adversaries	Associates
Rivals	Affiliates
Antagonists	Partners
Enemies	Friends
Hostiles	Comrades
Oppressing	Safeguarding
Intimidating	Diffusing
Threatening	Diverting
Frightening	Deterring
Bullying	Preventing
Menacing	Avoiding
Abusing	Withdrawing
Aggressing	Defending
Invading	Countering
Assaulting	Reciprocating
Attacking	Retaliating
Contending	Subduing
Fighting	Overcoming
Battling	Capturing
Controlling	Co-existing
Dominating	Equalizing
Subjugating	Balancing
Enslaving	Cooperating
Disabling	Collaborating
Injuring	Integrating
Killing	Unifying
Complexity	Simplicity
Dissatisfaction	Satisfaction
Greed	Sufficiency
Anger	Serenity
Unrest	Stillness
Hatred	Solitude
Fear	Silence

APPENDIX ONE
INTRAPERSONAL RELATIONSHIPS

For the purposes of this rendition of *The Art of War,* the intra-personal relationship constituents of the self of human beings are: 1) our True Tao-Nature or Higher Self and 2) our whole ego-Self which is the integration of 3) our executive ego-self and 4) our subordinate ego-selves or subpersonalities. The following table shows some of these constituents and their modes, characteristics, activities and relationships.

TRUE TAO-NATURE	EXECUTIVE EGO-SELF	MODE OF WHOLE EGO-SELF	NATURE OF SUBPERSONALITIES
Always openly and neutrally observing, witnessing and beholding the realities, activities and relations of our whole ego-Self	Mostly primarily, actively and fully balancing, integrating and uniting the realities, activities and relations of our sub-personalities	Being	Ego-Self Bound-Free Closed-Open
		Self	Unreal-Real Ill-Well Divided-Whole
		Thinking	Fact-Mytery False-True Unclear-Clear
		Feeling	Negative-Positive Bad-Good Down-Up
		Acting	Wrong-Right Difficult-Easy Failed-Successful
		Relating	Many-One Distant-Close Cold-Warm

Appendix Two
Interpersonal Relationships

For the purpose of this rendition of *The Art of War*, the interpersonal relationship constituents of the self of human beings are: 1) our True Tao-Nature or Higher self, 2) our whole ego-Self which is the integration of 3) our executive ego-self and 4) our subordinate ego-selves or subpersonalities and 5) other whole ego-Selves and their executive ego-self and subordinate ego-selves and our relationship with them. The following table shows some of these constituents and their modes, characteristics, activities and relationships.

True Tao-Nature	Executive Ego-Self	Mode of Whole Ego-Self	Other Whole Ego-Selves
Always openly and neutrally observing, witnessing and beholding the realities, activities and relations of our whole ego-Self	Mostly primarily, actively and fully balancing, integrating and uniting the realities, activities and relations of other whole ego-Selves	Co-existing	Included, Interconnected, Integrated
		Interrelating	Correlating, Interacting, Interchanging
		Harmonizing	Equalizing, According, Balancing
		Cooperating	Co-acting, Collaborating, Synergic
		Opposing	Competing, Rivaling, Vying
		Conflicting	Disputing, Arguing, Quarreling
		Warring	Contending, Fighting, Battling

NON-PERSONAL RELATIONSHIPS

For the purposes of this rendition of *The Art of War*, the non-personal relationship constituents of human beings are: 1) our True Tao-Nature or Higher Self, 2) our whole ego-Self which is the integration of 3) our executive ego-self and 4) our subordinate ego-selves or sub-personalities and 5) our ecological environment, natural resources, biodiverse species, indigenous peoples and plant medicines; their forms, conditions, situations and characteristics and our relationship to them. The following table shows some of these constituents and their modes, characteristics, activities and relationships.

True Tao-Nature	Executive Ego-Self	Mode of Whole Ego-Self	Our Natural Environment
Always openly and neutrally observing, witnessing and beholding the realities, activities and relations of our whole ego-Self	Mostly primarily, actively and fully balancing, integrating and uniting the realities, activities and relations of our outer environment	Co-existing	Ecosystems / Habitats / Biodiversity
		Interrelating	Guardianship / Custodianship / Stewardship
		Safeguarding	Protecting / Maintaining / Sustaining
		Harmonizing	Preserving / Reserving / Conserving
		Cooperating	Not Controlling / Not Dominating / Not Exploiting
		Opposing	Separate / Divided / Alienated
		Conflicting	Bad Soil / Bad Water / Bad Air
		Warring	Disasters / Calamities / Catastrophes

Appendix Four
Transpersonal Relationships

For the purposes of this rendition of *The Art of War*, the transpersonal relationship constituents of human beings are: 1) our True Tao-Nature or Higher Self, 2) our whole ego-Self which is the integration of 3) our executive ego-self and 4) our subordinate ego-selves or subpersonalities and 5) the resolution of oppositions and conflicts through, across and between various relationships. The following table shows some of these constituents and their modes, characteristics, activities and relationships.

TRUE TAO-NATURE	EXECUTIVE EGO-SELF	MODE OF WHOLE EGO-SELF	THE RESOLUTIONS
Always openly and neutrally observing, witnessing and beholding the realities, and relations of our whole ego-Self	Mostly primarily, actively and fully balancing, integrating and uniting the realities, and relations of our various resolutions	Connecting	Meeting / Engaging / Joining
		Interrelating	Interacting / Sharing / Exchanging
		Correlating	Complementing / Coordinating / Cooperating
		Harmonizing	Equalizing / Balancing / According
		Uniting	Integrating / Communing / Assimilating
		Not Separating	Not Dividing / Not Isolating / Not Alienating
		Not Conflicting	Not Opposing / Not Competing / Not Contending
		Non Eliminating	Not Negating / Not Destroying / Not Annihilating

APPENDIX FIVE
HISTORICAL INTERNATIONAL WARS

Dictionary definitions of 'War' are 'A state of opened and declared armed hostile conflict within and/or between different countries, nations, states and territories causing at least 1,000 deaths annually'. 'All conflicts and wars of which involve existential threats to humanity.'

A lengthy chronological listing of known historical wars, with mortality numbers greater than 25,000 human beings, from before 1000 CE until the present time can be found on the Wikipedia web-site. There reportedly have been over an estimated 14,500 wars world wide since before 1000 CE; 2,000 since the 1800s; 1,400 since the end of WW I in 1918 and 18-56 currently ongoing; only approximately 200 of which have historically involved less than 25,000 deaths.

Most of these wars have resulted in the deaths of hundreds of thousands of human beings who are combatants and casualties in conventional wars and which do not include civilian fatalities due to concomitant war-related famines, starvation, epidemics, atrocities and genocide.

The following lists some multi-national world-wide wars that 1) have been variously listed as 'struggles, clashes, skirmishes and riots; resistances, uprisings, rebellions, revolts and insurrections; coups, conspiracies and mutinies; campaigns, sieges, raids, incursions and invasions; conquests, takeovers and occupations and conflicts, feuds, crusades, battles, wars, civil wars, slaughters and massacres' and that 2) each have involved the deaths of countless millions of human beings, as noted:

475-221 BCE ❖ Chinese Warring States ❖ 1,500,000.

264-146 BCE ❖ Punic Wars ❖ 3,240,000 – 3,840,000.

66-136 CE ❖ Jewish-Roman Wars ❖ 1,520,000 – 3,100,000.

184-205 ❖ Chinese Yellow Turban Rebellion ❖ 3 – 7,000,000.

184-220 ❖ 3 Chinese Kingdom War ❖ 36 – 40,000,000.

629-1050 ❖ Arab-Byzantine Wars ❖ 2,000,000.

755-763 ❖ Chinese An Lushan Rebellion ❖ 13 – 36,000,000.

1095-1291 ❖ Crusades ❖ 1 – 3,000,000.

1206-1368 ❖ Mongol Invasions/Conquests ❖ 30 – 40,000,000.

1337-1453 ❖ 100 Years War ❖ 2,300,000 – 3,500,000.

1470-1574 ❖ Mediterranean War ❖ 900,000 – 1,000,000.

1499-1540 ❖ Spanish Conquest of Columbia ❖ 5 – 8,000,000.

1519-1530 ❖ Spanish Conquest of Mexico ❖ 8 – 10,000,000.

1519-1595 ❖ Spanish Conquest of Yucatan ❖ 1,460,000.

1533-1566 ❖ Spanish Conquest of the Inca Empire ❖ 8,400,000.

1562-1598 ❖ French Religion Wars ❖ 2 – 4,000,000.

1616-1683 ❖ Chinese Ming to Qing Dynasties ❖ 25,000,000.

1618-1648 ❖ 30 Years War ❖ 4 – 12,000,000.

1701-1714 ❖ War of Spanish Succession ❖ 400,000 – 1,250,000.

1756-1763 ❖ 7 Years War ❖ 86,000 – 1,400,000.

1771-1802 ❖ Chinese Tay Son Rebellion ❖ 1,200,000 – 2,000,000.

1803-1815 ❖ Napoleanic Wars ❖ 3,500,000 – 7,000,000.

1808-1833 ❖ Spanish-American War ❖ 600,000 – 1,200,000.

1850-1864 ❖ Chinese Taiping Rebellion ❖ 20 – 70,000,000.

1854-1856 ❖ Chinese Red Turban Rebellion ❖ 1,000,000.

1854-1873 ❖ Chinese Miao Rebellion ❖ 4,900,000.

1856-1873 ❖ Chinese Panthay Rebellion ❖ 890,000 – 1,000,000.

1861-1865 ❖ United States Civil War ❖ 650,000 – 1,000,000.

1862-1877 ❖ Chinese Dungan Revolt ❖ 8 – 20,000,000.

1910-1920 ❖ Mexican Revolution ❖ 1,000,000 – 3,500,000.

1914-1918 ❖ World War I ❖ 17 – 40,000,000.

1917-1922 ❖ Russian Civil War ❖ 7 – 12,000,000.

1927-1949 ❖ Chinese Civil War ❖ 8 – 11,700,000.

1936-1939 ❖ Spanish Civil War ❖ 500,000 – 1,000,000.

1937-1945 ❖ 2nd Sino-Japanese War ❖ 20 – 25,000,000.

1939-1945 ❖ World War II ❖ 80,000,000.

1946-1948 ❖ India Partition ❖ 200,000 – 2,000,000.

1947-1948 ❖ Indo-Pakistani War ❖ 200,700 – 2,000,000.

1950-1953 ❖ Korean War ❖ 3,500,000 – 4,500,000.

1954-1962 ❖ Algerian War ❖ 400,000 – 1,500,000.

1955-1975 ❖ Vietnam War ❖ 1,300,000 – 4,300,000.

1967-1970 -❖ Nigerian Civil War ❖ 1 – 3,000,000.

1971-1971 ❖ Bangladesh Liberation War ❖ 400,000 – 3,600,000.

1974-1991 ❖ Ethiopian Civil War ❖ 500,000 – 1,500,000.

1978-2021 ❖ Afghanistan Conflict ❖ 1,450,000 – 2,600,000.

1979-1989 ❖ Soviet-Afghan War ❖ 600,000 – 2,000,000.

1980-1988 ❖ Iran-Iraq War - 500,000 ❖ 1,500,000.

1983-2005 ❖ 2nd Sudanese Civil War ❖ 1 – 2,000,000.

1998-2003 ❖ 2nd Congo War ❖ 2,500,000 – 5,400,000.

2001-2021 ❖ War on Terror ❖ 272,000 – 1,260,000.

2003-2011 ❖ U.S.-Iraq Invasion and War ❖ 800,000 – 1,000,000.

2006-Now ❖ Mexican Drug War ❖ 200 – 400,000 +.

2014-Now ❖ War against the Islamic State ❖ 80,000 – 1,000,000 +.

2022-Now ❖ Russian-Ukraine Invasion and War ❖ 1,000,000 +.

2023-Now ❖ Israel-Hamas/Hezbollah War ❖ 40,000 and rising.

Most wars are started and conducted by autocratic leaders and governments for reasons of survival and expansion; for the acquisition of lands, territories, resources and populations and for economic, business and financial reasons that involve the control, compliance, domination, subjugation, submission, enslavement, exploitation, injury, killing and loss of countless human beings.

It is noteworthy that, historically, relatively far fewer disputes, conflicts and wars have occurred within and between countries and nations led by women in positions of political power.

Multi-millions of blessed human Beings, precious human Souls and sacred human Spirits have been, and are currently being, killed by military forces in major world-wide armed disputes, conflicts, fights, battles and wars at a reported rate of 2-3 per year, a current annual global cost of 18 trillion dollars and an average GDP of 13%.

What are real and true reasons for even starting any war? What is really and truly worth killing human beings for?

YOUR REFLECTIONS

NOTES

1. *The Art of War* can be synonymously identified as the Tao, Way or Method of War, the latter being the original way Sun Tzu's *The Art of War* was titled as *Sun Tzu's Warfare Methods*. The various meanings of 'Tao' are way, road and path; method; The Way and Ultimate Reality and the way in which the universe, Nature and all living beings and things are being, living, interrelating and going. Henceforth, the *Art of War* will be referred to as the Tao/Way/Art of War.

2. 'Conflict' is used generically to variously refer: 1) to frictions, irritations and aggravations; disagreements, controversies and disputes; arguments, quarrels and altercations; hostilities, contentions and confrontations; clashes, attacks and assaults; and fights, battles, combats and wars and 2) to militant encounters and engagements, relationships and interactions, campaigns and operations and conduct and activities.

'Opponent' is used generically to variously refer to:
Enemies, foes, fighters, combatants, battlers, strugglers and invaders.
Conquerers, vanquishers, subduers, dominators and subjugators.
Suppressors, repressors, oppressors, harassers and persecuters.
Aggressors, offenders, perpetrators, assailants, attackers and assaulters.
Lawless conspirators, secret planners, plotters, schemers and saboteurs.
Abusers, violaters, victimizers, objectifiers, 'other' – and 'thing'-makers.
Protesters, confronters, agitators, instigators, provokers and inciters.
Adversaries, antagonists, rivals, competitors, contenders and challengers.

And

Those 'others' who, in relation to human beings:

Murder, kill, rape, kidnap, abduct, capture, imprison, conceal and traffic.

Are predatory, rapacious, covetous, possessive and exploitive.

Are inhumane, savage, barbarous, vicious, merciless, fierce and noxious.

Traumatize, violate, molest, abuse, threaten, menace, intimidate, bully.

Frighten, terrify, terrorize, victimize, agonize, torment and torture.

Are toxic, brutal, cruel, sadistic, vengeful, revengeful and vindictive.

Accost, harm, hurt, bruise, wound, injure, maim, cripple and disable.

Are combative, belligerent, defiant, demanding, nagging and pestering.

Override, overpower, overwhelm, overcome, overthrow and overrule.

Are restricting, constricting, restraining, constraining and confining.

Control, manipulate, force, coerce, dominate, subjugate and enslave.

Persuade, induce, conduce, bribe, seduce, sway, debauch and corrupt.

Contend, wrangle, altercate, struggle, tangle, wrestle and grapple.

Are defeating, besting and triumphing and gloating over winning.

Dissent, disagree, dispute, debate, controvert, argue, quarrel and squabble.

Push, shove, jab, hit, strike, punch, kick, batter, pound on and beat up.

Frustrate, irritate, irk, annoy, harass, aggravate, rile, disturb, agitate.

Instigate, incite, foment, abet, provoke, stir up, enrage and
  infuriate.

Mistreat, mishandle, mismanage, place at risk, endanger and
  imperil.

Inflict and cause upset, difficulty, trouble, frustration and
  anger.

Afflict and cause pain, distress, grief, anguish, misery and
  suffering.

Dishonor, disrespect, disgrace, disregard, ignore, slight and
  neglect.

Devalue, debase, vitiate, demean, degrade, derogate and
  deplore.

Doubt, question, disbelieve, villify, defame, malign and
  slander..

Depreciate, deprecate, belittle, shame, denigrate and disparage.

Ridicule, make fun of, deride, mock, rally, taunt, twit and
  mimic.

Humiliate, abase, mortify, embarrass, affront, insult and offend.

Disapprove, disavow, disaffirm, disprove, discredit and dismiss.

Diminish, disadvantage, handicap, deprive, impoverish, and
  dispirit.

Suspect, distrust, judge, blame, accuse, incriminate and
  condemn.

Invalidate, disqualify, reject, rebuff, repudiate, denounce and
  disown.

Are unkind, uncivil, discourteous, inconsiderate, intrusive and
  invasive.

Are condescending, belittling, disparaging, mocking, deriding,
  ridiculing.

Are defensive, resistant, uncooperative, stubborn and
  obstreperous.

Are disobedient, challenging, defiant, rebellious, radical and
  insurgent,

Are belligerent, bellicose, contentious, pugnacious and combative.

Are toying with, playing with, gaming, cheating, using and
    exploiting.

Lie, steal, rob, defraud, betray, deceive, fool, scam, con, trick
    and dupe.

Severely criticize, reprehend, rebuke, scold, reprove and
    reprimand.

Admonish, censure, chastise, castigate, harshly discipline and
    punish.

Cajole, lecture, hassle, harangue, plague, perturb, outrage and
    inflame.

Dehumanize, depersonalize, marginalize, deprive and
    disenfranchise.

Stigmatize, pathologize, patronize, infantilize, demoralize and
    dispirit.

Disempower, weaken, disable, undermine, devitalize, enervate
    and unnerve.

Discriminate, segregate, isolate, disadvantage, profile and
    target.

Alienate, estrange, ostracize, exclude, ban, exile, banish and
    expel.

Negate, nullify, eliminate, exterminate, eradicate and
    annihilate.

    And

    Those 'others' who, in relation to human beings, have:

    Stereotypes, prejudice, bias and favoritism.

    Hatred, animosity, antipathy and enmity.

    Distaste, dislike, aversion and repulsion.

    Disgust, despisement, disdain and scorn.

    Loathing, repugnance and abhorrence.

    Revulsion, contempt and detestation.

    Hostility, ire, anger, rage, wrath and fury.

    Malignity, maliciousness and malevolence.

    Grudges, resentment, ill-will and spite.

    Any and all other 'negative' feelings and actions.

When reading through this rendition of *The Art of War*, simply remember that:

1) 'human beings' are referred to as our whole 'ego-Self' which is the co-existing integration of our True Tao-Nature, our executive ego-self and our subordinate ego-selves within our consciousness, conscious awareness and conscious experience.

2) 'wars, battles and fights', etc. are generally referred to as 'conflicts'; intrapersonal ones within ourselves, interpersonal ones within our relationships with opposing fellow human beings and non-personal ones within our relationships with environments; that are occurring within our consciousness, conscious awareness and conscious experience.

3) 'adversaries, antagonists, aggressors, attackers', etc.; within either intrapersonal or interpersonal conflicts within our consciousness, conscious awareness and conscious experience; are generally referred to as 'opponents' or 'the opposition'.

4) our True Tao-Nature is our Tao-Self, Higher and Deeper Self and Pure Consciousness.

5) our executive ego-self uniquely integrates, harmoniously organizes, selectively activates, functionally mobilizes, intentionally directs and appropriately utilizes various subordinate ego-selves or subpersonalities in the service of safeguarding and insuring the reality and presence of our True Tao-Nature and of maintaining and sustaining the viability and integrity of our whole ego-Self within our consciousness, conscious awareness and conscious experience.

6) our subordinate ego-selves are harmoniously integrated and cooperatively united with our executive ego-self and with each other or are opposing, competing and conflicting with each other; defending, advancing and aggressing against each other and/or overpowering, overcoming and defeating each other within our consciousness, conscious awareness and conscious experience.

7) all intrapersonal, interpersonal and non-personal selves and opponents; structures and dynamics; relationships and

activities, harmonics and conflicts, etc. are concepts within our consciousness, conscious awareness and conscious experience.

3. An early characterization of the intrapersonal conflict, struggle and battle between body and Spirit within our consciousness, conscious awareness and conscious experience, i.e., our human Soul, is found in the *Bhagavad Gita/Song of the Lord/God,* the central portion of the Hindu epic *Mahabarata* (c. 5th-2nd Centuries BCE).

Here there is a dialogue between: 1) Lord Krishna; the incarnation of the god Vishnu, the preserver and sustainer, and the supreme spiritual Self, and 2) Arjuna, the embodied warrior prince concerning the latter's need to fulfill his duty (karma yoga) as a warrior and to battle even the relatives and friends whom he realizes and knows (jnana yoga), whom he loves and is devoted to (bhakti yoga) and with whom he shares an ultimate union (raja yoga).

Metaphorically, the epic characterizes the intrapersonal and intrapsychic conflicts between 1) ultimate True Tao-Nature and the whole ego-Self of human beings and 2) its executive ego-self and subordinate ego-selves within the field of human consciousness, conscious awareness and conscious experience.

Three examples of topographical intrapsychic and intrapersonal structures and dynamics are those of Sigmund Freud's Psychoanalysis, Carl Jung's Analytic Psychology and Roberto Assagioli's Psychosynthesis. Each approach describes a fundamental intrapsychic structural and dynamic duality of conscious and unconscious and various intrapsychic dualities that can either be in conflict with each other or ultimately integrated and realized.

In Psychoanalysis; 1) id/primitive, instinctual and impulsive needs; 2) ego/rational organizing, mediating and controlling functions and 3) superego/social moral and normative standards of conscience are either conflictingly or harmoniously co-present, co-valent and co-variant in the unconscious/id, conscious/

ego and overconscious/superego of the psyche of human beings.

In Analytic Psychology; there are numerous structural, binary antithetical and conjunctive opposites in the conscious and the personal and collective unconscious of the psyche; e.g., 1) mother-father, Electra-Oedipal, inferiority-superiority complexes; 2) shadow-persona, anima-animus, senex/wise old man-puer aeturnus/eternal child and the Self archetypes; 3) dreams-conscious experience, intuition and imagination-intellect; extroversion-introversion personality traits and 4) archetypes such as creator-devil, mother-father, wise woman-man, sage-ruler, hermit-recluse, hero-warrior, child-orphan, lover-caregiver, magician-trickster, jester-joker, rebel-outlaw, wild-madman et al. Opposites are integrated via synchronicities and the transcendent functioning of the individuation process leading to realizing the inner psychic center, totality and wholeness of the Self.

In Psychosynthesis; the unifying essential, transpersonal supersonscious, spiritual Being and Higher Self co-exist along with various subpersonalities within the lower, middle and higher unconscious and conscious ego of the psyche of human beings. The subpersonalities either are conflictingly opposed to each other or are disidentified from or harmoniously synthesized and integrated, through the universal Will, Higher transpersonal Self and superconscious.

Subpersonalities are complexes and traits and images and concepts of the ego-self and are, e.g., the inner child and some that are variously seeking, adventuring, philosophizing, aesthetic, teaching, guiding, protecting, responsible, organizing, controlling, strategizing, perfectionistic, criticizing, challenging, rebelling, joking, wild, sabotaging, victimized, addicted, worrying et al.

Subpersonalities also include bipolar opposites such as love-hate, sympathy-antipathy, optimism-pessimism et al that are either balanced and synthesized or disidentified from in order to experience the primacy, unity and centrality of superconsciousness and the deeper and higher transpersonal Self.

Also, psychologically; regarding the general existence of divided and antithetical intrapsychic personalities in the psyches of some individual human beings; 1) 'doppelgangers' or 'doubles' are present; 2) 'split' personalities exist, e,g., Jekyll and Hyde, good-me and bad-me and 3) any number of 'multiple' personalities and identities are present that are commonly dissociated from each other with no one personality that is usually, necessarily or essentially primary, central, dominant, unifying or integrating.

4. Interpersonal relationships, unions and harmonies and oppositions and conflicts within our consciousness, conscious experience and conscious awareness also include those between ourselves and various animals and any other of Mother Nature's creatures. Such beings are observed, appreciated and admired in the wild; protected in sanctuaries, preserves, reservations and zoos and, as pets, are cared for in animal shelters, rescue adoption facilities and private homes.

Such animal-beings are related to in much the same way that we, as human beings, intimately relate to our fellow human beings and with the same kind, degree and extent of value, respect, love, care, satisfaction, companionship and happiness.

Canine-beings are invaluable and crucial resources used: 1) as guide-dogs by blind human beings, 2) as search and rescue and cadaver dogs for finding human beings buried in wreckages and rubble and missing from home and/or murdered and for finding and rooting out enemy personnel in wars, 3) as support dogs for apprehending criminals in police work, 4) as detective dogs for sniffing out trafficked drugs and 5) as faithful and helpful companion dogs for children and especially for terminal, disabled, ill, elderly, lonely and otherwise needful human beings.

Unfortunately, our interpersonal relationships with too many of our animal-beings and -friends include, e.g., trophy hunting and poaching; illegal whaling and overfishing; domesticating, industrially farming, confining, overcrowding and fattening them; using hormones and antibiotics and slaughtering, butchering, dressing and saran-wrapping parts of them for sale in meat-markets for carnivorous human consumption.

5. Non-personal relationships within our consciousness, conscious awareness and conscious experience are those with Nature and our planetary and ecological environment. They are, of course, also intensely personal ones. There is no war on Nature; no overcoming, conquering, vanquishing and defeating It; no subduing, mastering, dominating and subjugating It; no being victorious and triumphant over It and no establishing of the supremacy and dominion of human nature and human beings over It.

There is committed planetary custodianship, stewardship and guardianship and enlightened global environmental and ecological consciousness and action. There is wise reservation of wilderness lands and rain forests, conservation of wildlife and natural resources and management and enhancement of ecosystems and environments. There is compassionate preservation of species biodiversity and animal habitats that does not involve their destruction, degradation and resultant endangerment and loss.

6. In various translations of *The Art of War*, direct/Cheng/Zheng and indirect/Ch'i/Qi strategic tactics are respectively rendered as: 1) orthodox, conventional, regular, ordinary, overt, direct, formed, solid, planned, fixed, set and expected and 2) unorthodox, unconventional, irregular, extraordinary, covert, indirect, unformed, insubstantial, spontaneous, fluid, flexible and surprising.

7. The following are not discrete and/or hierarchical categories, but may have some heuristic value when considering the nature, modes of being, qualities, characteristics, functions, activities and endings of thinking, feeling, acting, being and relating and conflicts and wars.

Conclusion – close, end.
Mind and Psyche – mental, cognitive, thinking and knowing.
Not abstractions, concepts, ideas and thoughts and ignorance, defilements and delusions.
Epistemology and Logic. Knowledge and reasoning.
Awakening and realization, pure consciousness and empty awareness and enlightenment and wisdom.
The integrative resolution of intrapersonal conflicts.
The harmonious integration of subordinate ego-selves.

Completion – fulfill, whole, perfect.
Heart and Soul – emotional, affective, feeling and having.
Not objects, things, goods and possessions and desires, afflictions and attachments.
Aesthetics and axiology. Beauty and values.
Dedication and compassion, Vital energy and transformation and illumination and radiant beauty.
The unopposed resolution of interpersonal conflicts.
The peaceful joining with an opposing whole ego-Self.

Culmination – crown, highest point.
Will and Soma – volitional, conative, acting and doing.
Not intentions, purposes, deeds and feats and obstructions, fetters and errors.
Ethics and politics. Morality and government.
Cultivating and practicing, liberating and actualizing and peacefulness and skillful means.
The active resolution of non-personal conflicts.
The unified co-existing of human beings and ecosystems.

Consummation – highest degree, complete, perfect.

Being and Socius – ontological, integral, relational and communal.

Not ego-illusion, egocentricity, separation, division, opposition, conflict and 'others'.

Metaphysics and ontology. Transcendence, co-being and inter-being.

Spiritual identity, True Tao-Nature and Higher and Deeper Tao-Self.

Fellow human beings, kindred Souls, Wayfaring companions and sojourning pilgrims.

Acceptance and inclusion, appreciation and gratitude, empathy, compassion and intimacy.

Generosity and forgiveness, peacefulness and happiness, freedom and fulfillment.

The mediating resolution of intrapersonal, interpersonal and non-personal conflicts.

The integral Way of transpersonal spiritual and soulful human being and interrelations.

Conflict and war
between and within.
No human Souls begin.
No human beings win.

Conflict and war
within and between.
All human Souls refuse.
All human beings lose.

END NOTE

Living ethically, morally, civilly, respectfully, truthfully, responsibly, trustingly, faithfully, kindly, peacefully, freely, happily, humbly, intimately, gratefully and courageously are; 1) natural components of purely and simply, sheerly and utterly, being awakened, realized, enlightened, transformed, liberated, original, universal and inborn True Nature in and as itself rather than; 2) intentionally, unnaturally, strategically and deviously trying to be and to live in such ways for the sole purpose of obtaining some benefit, reward, merit, good karma, grace, dispensation, absolution, expiation, redemption, deliverance, salvation, etc..

Being a real, true, good and right human being and living and sharing a real, true, good and right human life of dignity, integrity, wisdom, freedom, happiness, compassion, friendliness, peacefulness and service beyond dishonor, dishonesty, ignorance, bondage, suffering, judgment, hatred, conflict and oppression are natural consequences of not completely identifying with our socially conditioned, fictitious, fabricated, artificial, illusional and dualistic ego-self and being the awakened, enlightened, transformed and liberated original and inborn True Nature who we absolutely, ultimately, essentially and universally *are*.

Such a Way of being occurs through cultivating and practicing being enlightened True Nature in the ordinary experiences of everyday living wisely, compassionately and skillfully with fellow human beings, kindred souls and spiritual sisters and brothers including those who are preoccupied with and addicted to money, power, status and success; sex and pornography; drugs, alcohol and medications; technology, smart-phones, electronic gadgets and video games; social media and virtual reality; gambling, vicarious entertainment and spectator sports; competing, striving and winning and aggression, fighting, violence and warfare.

Overcoming, awakening, enlightening, transforming and liberating our socially conditioned and determined, contrived, devised and conventionally accepted concept of an ego-self from its unreality and illusion; autonomy and hegemony; ignorance and delusion; desires and attachments; habits and errors; separations and divisions; dualities and oppositions and its controlling and defending; opens the Way to becoming, being, experiencing, living, enjoying and sharing ourselves as one, original, universal and inborn True Nature that is absolutely, ultimately, essentially, naturally, inherently and eternally conflict-free and warring-free.

It is insufficient to simply either cynically, realistically or compassionately accept the apparent reality that it is an unfortunate tragic part of human nature and the human condition that there can or will be no end to human beings killing fellow human beings; either justifiably, randomly or because of various assumed differences, disagreements, disputes, antagonisms, enmities, hostilities, animosities, conflicts, wars, etc..

What is going to prevent and stop human beings from killing one another?

1) Sincere prayers to an all-powerful intervening deity or divinity?

2) The pandemic-level occurrence of all-powerful planetary shifts?

3) The overwhelming occurrences of all-powerful natural events?

4) The commanding appearance of an all-powerful salvific being?

5) A spiritual consortium of all-powerful enlightened human beings?

6) The egalitarian wisdom of all-powerful international leaders?

7) Mandatory edicts by all-powerful governmental bodies?

8) Severe punishments for all-powerful illegal violations?

9) The effective exercise of our all-powerful collective conscience?

10) The revolutionary uprising of all-powerful organized protesters?

11) The awesome visitation by all-powerful extraterrestrial beings?

12) An interplanetary attack by all-powerful extraterrestrial beings?

13) The incredible manifestation of all-powerful parallel universes?

14) The feared possibility of all-powerful nuclear annihilation?

15) The disastrous effects of all-powerful global degradation?

16) The economic reality of all-powerful epidemic poverty?

17) The pervasive awareness of our all-powerful human suffering?

18) The human deaths of an all-powerful exceeded critical mass?

19) The loss and intense pain of all-powerful grieving loved ones?

20) The compelling impact of all-powerful uncontrolled events?

21) A cataclysmic apocalypse of all-powerful human existence?

22) A brilliant manifestation of all-powerful cosmic Light?

23) A collective awakening to our all-powerful truthful Reality?

24) A collective actualizing of our all-powerful universal Purpose?

25) The arrival of all-powerful Maitreya Buddha or second coming of all-powerful Jesus Christ either in the world or their divine Presence, bright Light, true Law, kind Love, sacred Lore and beneficent Labor deep within our human hearts and minds?

26) Or sadly, perhaps none of these in our lifetime?

So far, the prevailing occurrence of world-wide viral pandemics and international wars have not served to prevent, reduce, limit and/or stop the mounting deaths and killings of human beings.

Too many sacred human souls and precious human beings

are being inhumanly, inhumanely and impersonally regarded as data, property, assets and 'packages'; interchangeable, replaceable and disposable 'tools' and commodities; objects, things, anonymous ciphers and non-entities; clients, customers, consumers and 'deals'; targets, 'marks', prey, 'johns' and victims; competitors, rivals, opponents and adversaries; antagonists, hostiles, threats, foes and enemies; numbers, statistics, 'units', body-counts and collateral damage, etc..

Killing our fellow human beings, kindred sisters and brothers and sacred human Souls and Spirit has absolutely no place in being, living and sharing; 1) our awakened, illuminated, enlightened, realized, liberated and 2) our absolute, ultimate, essential, one, universal, humane and inborn True Tao-Nature.

Survival, opposition, acquisition, competition, conflict, aggression, violence and war do not have to be accepted; 1) as inescapable, inevitable and untranscendable primal, instinctual, biological and 'hard-wired' given realities of our human nature and condition and 2) as a significant and determining part of our pure, clear and open consciousness and our real and true conscious human co-being, co-existing, living and interrelating!

While we may not be able to be in positions to directly make governmental decisions to not declare and engage in wars; we can influence and affect them by realizing the inhumanity, insanity and futility of war and by giving collective voice to active social protests. And, more possible, we can not engage in and/or quickly resolve our own struggles, disputes, conflicts, fights, battles and wars within ourselves and with and between our fellow human beings.

Given the choices of and in our lives, are we going to be WORRIERS or WARRIORS? Exchanging the second letters of the two words, i.e., the letter 'o' and the letter 'a', characterizes the difference:

Worrier = Onlooking, observing, objecting, opposing, obsessing thoughtfully and pensively.

Warrior = Awakening, abhoring, acting, asserting and averting courageously and bravely.

Changing the letters of words is easy. Changing and committing ourselves is more difficult.

Instead of inhabiting and surviving on
a few small endangered islands of peace
amid a large turbulent sea of conflict and war;
hope, pray and actively work for creating
a world of large thriving islands of peace
amid a small ebbing ocean of conflict and war.

Identify with the true
universal inner essence
of human beings and not
objectify their particular
individual outer form,
appearance and actions
and you won't have
'friends' or 'enemies',
'harmony' or 'conflict'
and 'peace' or 'war',
all of which are just
the invented names
of beings and things.

An Ode To War

WAR:
Being Spirit living as our Sacred core,
there is absolutely no place for war.
What is war really and truly for?

THE REASON:
Things that we admire and adore?
Needing, desiring, wanting more
to add to our heaped-up store?

THE ENEMY:
Our differences are right at the fore.
Our ideologies, beliefs and folklore.
Only those mine and not those your.

THE DUALITIES:
I and me, me and you, an either-or.
We and us and them, a neither-nor.
Dualities and conflicts rise and soar.

THE BETRAYAL:
Unkept promises that we swore.
A life dreamed of as never before.
Reality withered like an old whore.

THE DECISION:
Faced with an old rusty squeaky door.
Offered a chalice with nothing to pour.
Dropped upon a cold dusty creaky floor.

THE BATTLE:
A fearsome challenging enemy roar.
An open wound and painful sore.
An unnecessary settling of the score.

THE RESULT:
All of the vitality, blood and gore.
Oozing, draining out from every pore.
The fabric of human decency is tore.

THE CHRONICLE:
A futile endeavor, a ship with no oar.
Its written history a repetitious bore.
A sleeping autocrat's midnight snore.

THE RECOVERY:
Damaged human Souls to restore.
The unscathed human Spirit's spore.
Destined to reach the other shore.

YOUR REFLECTIONS

Glossary[*]

Chan/Zhan – war, warfare, combat, battle, fight.

Ch'e/Che – retreat.

Cheng/Zheng – fight, conflict, contend.

Cheng/Zheng – orthodox, conventional, planned, fixed, overt, direct, straightforward tactics.

Chi/Ji – plan, calculate, intention.

Chi/Zhi – foreknowledge.

Ch'i/Qi – unorthodox, unconventional, surprise, flexible, covert, indirect, devious tactics.

Ch'i/Qi – vital energy and force.

Chiang Chun/Jiang Jun – military general.

Ch'iang/Qiang – strong.

Chien/Jian – between, among.

Ch'ih/Chi – ruler.

Chin/Jin – advance.

Chiu/Jiu – nine, numerous.

Chu/Ju – plan.

Chun/Jun – arms, army, armed forces, military, martial.

Fa/Fa – method, art.

Fan/Fan – oppose.

Ho/He – together, with, harmony.

Hsien/Xian – danger.

Hsing/Xing – form, shape, figure, configuration, disposition (of forces).

Hsing/Xing – go, travel, act, carry out.

Hsu/Xu – empty, weak.

Huo/Huo – fire, flame.

I/Yi – art.

Kung/Gong – attack, assault.

Kung/Gong – civil, public.

Li/Li – advantage, benefit.

Ling/Ling – leader.

* The Chinese words are respectively those of the Wade-Giles and Pin-Yin systems.

Man/Man – full.

Mao/Mao – offense.

Mien/Mien – confront.

Mou/Mou – scheme, stratagem.

Pai/Bai – defeat.

Pien/Bian – transform, change.

P'ien/Pian – deceive.

Ping/Bing – weapon, soldier, military, troops, war.

Pu Li/Bu Li – no advantage, disadvantage.

Sheng/Sheng – victory.

Shih/Shi – configuration, situation, circumstances.

Shih/Shi – strategic power, potential energy, momentum.

Shih/Shi – substantial, solid.

Shou/Shou – defend, guard.

Sun/Sun – grandchild.

Tao/Dao – road, path, way, method, The Way, Ultimate Reality.

Te/De – virtue, the individualized power, virtuosity of Tao/Dao.

Ti/Di – earth, ground, terrain, place, situation.

Ti/Di – enemy.

T'ien/Tian – heaven.

Ts'e/Ce – strategy, plan.

Tzu/Zi – child, son, philosopher, master

Wei/Wei – position, location.

Wen/Wen – cultural, civil, peaceful.

Wu Ch'i/Qi – weapon, military utensil.

Wu-Wei – no, not, nothing doing; unforced and effortless action.

Wu/Wu – martial, military, warring.

Yang/Yang – sunny, light, open, overt, active, assertive, hot, hard, strong, high, full, et al.

Yin/Yin – shady, dark, hidden, covert, passive, receptive, cold, soft, weak, deep, empty, et al.

Yung/Yong – use, employ.

REFERENCES

Ames, Roger. Sun Tzu: *The Art of War*. New York: Ballantine Books. 1993.

Cleary, Thomas. Sun Tzu: *The Art of War*. Boulder, Colorado: Shambhala Publications, Inc.. 1988 and 2019.

Fenn, C. H.. *The Five Thousand Dictionary: Chinese-English*. Cambridge, Massachusetts: Harvard University Press. 1976.

Gagliardi, Gary. *The Art of War Plus the Ancient Chinese Revealed*. Seattle, Washington: Clearbridge Publishing. 2004.

Giles, Lionel. *Sun Tzu on the Art of War: The Oldest Military Treatise in the World*. London: Luzak & Co.. 1910.

Giles, Lionel. *The Art of War: Sun Tzu*. eBook. Las Vegas, Nevada. 2015.

Giles, Lionel. Sun Tzu: *The Art of War*. Bilingual Chinese and English Text. Rutland, Vermont: Tuttle Publishing. 2016.

Giles, Lionel. *The Art of War: Sun Tzu*. Deluxe Collection Edition. Bob Sutton/The Project Gutenberg EText. North Chelmsford, Massachusetts: Amazon Printing. 2021.

Giles, Lionel. *Sun Tzu: The Art of War. The Classic Guide to Strategy*. New York: St. Martin's Publishing Group. 2022.

Griffith, Samuel B.. *Sun Tzu: The Art of War*. Oxford: Oxford University Press. 1963.

Huynh, Thomas. *The Art of War: Spirituality for Conflict*. Woodstock, Vermont: Skylight Paths Publishing. 2008.

Ivanhoe, Philip J.. *Master Sun's Art of War*. Indianapolis: Hackett Publishing Co. Inc.. 2011.

Kaufman, Steven F.. *Sun Tzu: The Art of War*. Rutland, Vermont: Tuttle Publishing. 2021.

Li Dong. *Concise Chinese Dictionary: Chinese-English/English-Chinese*. Rutland, Vermont: Tuttle Publishing. 2015.

Mair, Victor H.. *The Art of War: Sun Zi's Military Methods.* New York: Columbia Univrsity Press. 2007.

Mathews, R.H.. *Mathew's Chinese-English Dictionary.* Cambridge, Massachusetts: Harvard University Press. 1943.

McNaughton, William and Lee Ying. *Reading and Writing Chinese Characters: Traditional Character Edition.* Rutland, Vermont: Tuttle Publishing. 1999.

Minford, John. *The Art of War: Sun Tzu.* New York: Penguin Books. 2002.

Nylan, Michael. *Sun Tzu: The Art of War.* New York: W.W. Norton & Company, Inc.. 2020.

Pepper, Jeff and Xiao Hui Wang. *The Art of War: A Step-by-Step Translation.* Verona Pennsylvania: Imagine Press. 2019-2021.

Quanyu Huang, Tong Chen and Kuangyan Huang. *McGraw-Hill's Chinese Dictionary and Guide to 20,000 Essential Words.* New York: McGraw-Hill. 2010.

Sawyer, Ralph D.. *Sun Tzu: The Art of War.* New York: Basic Books. 1994.

Webster's New Collegiate Dictionary. Springfield, Massachusetts: G. & C. Merriam Company. 1979.

Wieger, L.. *Chinese Characters: Their Origin, Etymology, History, Classification and Signification.* L. Davrout (Tr.). New York: Dover Publications. 1965.

Wilder, G.D. and Ingram, J.H.. *Analysis of Chinese Characters. New York: Dover Publications. 1974.*

Wing, R.L.. *The Art of Strategy: A New Translation of Sun Tzu's Classic, The Art of War.* New York: Doubleday. 1988.

Readers interested in comparing and contrasting related considerations of the nature, philosophies, politics, theories, principles, objectives, planning, strategies, tactics and conduct of offensive and defensive military armed conflicts, wars and warfare are referred to Kautilya/Chanakya's *Arthashastra* (2nd C. BCE), Niccolo Machiavelli's *The Art of War* (1521 CE), Carl Von Clausewitz's *On War* (1832 CE) Baron Antoine-Henri de Jomini's *The Art of War* (1838 CE), Leo Tolstoy's *War and Peace* (1865-1869 CE) and Julius Evola's T*he Metaphysics of War* (1935-1950 CE) as well as to relevant Google searches and You Tube videos.

References focusing upon the ethics, morality and legality of war; the reasons for war; the conditions for how wars should be fought and the differences between pacifism, idealism, liberalism, constructivism, realism and just war theory can be found in philosophical writings throughout the Ancient (~6th-5th C. CE), Dark/Middle Ages/Medieval (~5th-15th C. CE), Renaissance (~mid 15th-early 17th C. CE), Early Modern (~16th-18th C. CE), Enlightenment (~17th-18 C. CE) Modern (1650-1900 CE) and Contemporary/Post-Modern (1900 CE-present) time periods. The theory of 'just wars' can be found specifically in the writings of Aristotle (384-322 BCE), St. Augustine (354-430 CE) and St. Thomas Aquinas (1491-1556 CE) as well as through relevant Google searches and in You Tube videos.

ABOUT THE AUTHOR

I am a former Associate Professor of Integral Counseling Psychology and a retired Licensed Marriage Family Therapist after; 1) enjoying a fifty year long professional career educating, training, supervising and mentoring graduate school counseling trainees and psychology interns and 2) conducting an individual psychotherapy and counseling private practice for children, adolescents and adults and individuals, couples and families.

I have studied with prominent philosophy and psychology educators and trained with master psychotherapists and counselors in the fields of Phenomenology and Existential, Humanistic and Transpersonal Psychology and Psychotherapy. I have worked as a clinical psychologist in a wide variety of inpatient hospitals, outpatient clinics and group and individual private practices and have held clinical directorship positions in several community counseling centers and substance abuse treatment centers.

I have experienced principal teachers, teachings, Masters and meditation practices in the spiritual traditions of Hinduism; Theravada, Mahayana, Vajrayana, Ekayana, Ch'an and Zen Buddhism; Taoism and Zen. I have participated in small group shamanic medicine circles and experienced consciousness awakenings and openings and the transformative healing power of various natural plants and psychotropic substances.

I have written five other books integrating the spiritual and practical wisdom of Taoism and Zen:

1) A rendition of Lao Tzu's *Tao Te Ching* with psychotherapeutic commentary.

2) A rendition of Chuang Tzu's *Seven Interior Records* with psychotherapeutic commentary.

3) A rendition of Lieh Tzu's *The Nature of Real Living* with psychotherapeutic commentary.

4) A rendition of Lao Tzu's *Tao Te Ching* with Soul-journeying commentary.

5) A rendition of *The Ten Ox-Herding Pictures* with psychospiritual commentary.

My former early life was enjoyed in intimate relationship with Great Mother Nature and Her creatures and living close to river, forest and lake natural settings. My current later life is being enjoyed living in a wooded natural setting beside a seasonal creek and close to the Pacific Ocean. Thus, a full circle of being and living is coming to completion.

Your Final Reflections

Milton Keynes UK
Ingram Content Group UK Ltd.
UKHW021112031224
452078UK00010B/849

9 781587 906923